Michael Jay Katz

From Research to Manuscript

A Guide to Scientific Writing

 Springer

Michael Jay Katz
Case Western Reserve University,
Cleveland OH
U.S.A.
mjk8@case.edu

ISBN 978-1-4020-9466-8   e-ISBN 978-1-4020-9467-5

Library of Congress Control Number: 2008940867

Printed on acid-free paper

springer.com

Acknowledgements

I thank journal editors Olaf Andersen, John Bennett, Lynn Enquist, David Fastovsky, Robert Genco, William Giannobile, Kathryn Harden, Theodore Harman, William Koros, James Olds, David Rosenbaum, Catharine Ross, and George Schatz for suggesting specific well-written scientific papers. Permissions to quote from their papers were granted by Daniela Berg, David Fastovsky, Jacqueline Geraudie, David Rosenbaum, and Benjamin Widom.

Journal permissions for quotations

- Augspurger et al. 2005. *J Nutr* 135: 1712–1717 [copyright 2005, permission from the American Society for Nutritional Sciences].
- Bohm et al. 2005. *Leukemia Res* 29: 609–615 [copyright 2005, permission from Elsevier].
- Borgens et al. 2004. *J Neurosci Res* 76: 141–154 [copyright 2004, permission from Wiley-Liss, a subsidiary of Wiley].
- Fastovsky and Sheehan. 2005. *GSA Today* 15: 4–10 [permission from the Geological Society of America].
- Gapski et al. 2004. *J Periodontol* 75: 441–452 [permission from the American Academy of Periodontology].
- Glaunsinger and Ganem. 2004. *J Exp Med* 200: 391–398 [copyright 2004, permission from The Rockefeller University Press].
- Milner et al. 1968. *Science* 161: 184–186 [copyright 1968, permission from AAAS].
- Readinger and Mohney. 2005. *J Electronic Mater* 34: 375–381 [permission from the Minerals, Metals & Materials Society].
- Sugimori et al. 1994. *Biol Bull* 187: 300–303 [permission from The Biological Bulletin].
- Sundar and Widom. 1987. *J Phys Chem* 91: 4802–4809 [copyright 1987, permission from the American Chemical Society].
- Williams CM. 1961. *Biol Bull* 121: 572–585 [permission from The Biological Bulletin].

Contents

Introduction

Observations *Plus* Recipes

It has been said that science is the orderly collection of facts about the natural world. Scientists, however, are wary of using the word 'fact.' 'Fact' has the feeling of absoluteness and universality, whereas scientific observations are neither absolute nor universal.

For example, 'children have 20 deciduous [baby] teeth' is an observation about the real world, but scientists would not call it a fact. Some children have fewer deciduous teeth, and some have more. Even those children who have exactly 20 deciduous teeth use the full set during only a part of their childhood. When they are babies and toddlers, children have less than 20 visible teeth, and as they grow older, children begin to loose their deciduous teeth, which are then replaced by permanent teeth.

'Children have 20 deciduous [baby] teeth' is not even a complete scientific statement. For one thing, the statement 'children have 20 deciduous teeth' does not tell us what we mean by 'teeth.' When we say "teeth," do we mean only those that can seen be with the unaided eye, or do we also include the hidden, unerupted teeth?

An observation such as 'children have 20 deciduous teeth' is not a fact, and, by itself, it is not acceptable as a scientific statement until its terms are explained: scientifically, 'children have 20 deciduous teeth' must be accompanied by definitions and qualifiers. The standard way to put science into a statement is to define the statement's meaning operationally. Instead of attempting a purely verbal definition of 'teeth,' for instance, scientists define it by the procedure—the recipe—that has been used when making the observations about teeth.

In science, 'children have 20 deciduous teeth' is neither universal nor abstract. It is a record of the result of following a specific recipe, and the statement is scientific only when we include the recipe that was used. For 'children have 20 deciduous teeth,' one appropriate recipe would be: "I looked in the mouths of 25 five-year-old boys and 25 five-year-old girls in the Garden Day Nursery School in Cleveland, OH, on Monday, May 24, 2008, and I found that 23 of the boys and 25 of the girls had 20 visible teeth."

A meaningful scientific statement includes an observation and its recipe, and the standard form for recording meaningful scientific statements is the scientific research paper.

Writing a Scientific Research Paper

Science is the orderly collection of scientific records—i.e., observations about the natural world made via well-defined procedures—and scientific records are archived in a standardized form, the scientific research paper. A research project has not contributed to science until its results have been reported in a standard paper, the observations in which are accompanied by complete recipes. Therefore, to be a contributing scientist, you must write scientific papers.

This book contains my advice and thoughts about writing a scientific research paper. My basic hard-won realization is that writing a good scientific paper takes time. On the other hand, I have found that the writing will seem endless if you begin with the title and slog straight through to the last reference. This approach is difficult, wearing, and inefficient. There is a much more effective way to write.

I suggest that you write your paper from the inside out. Begin with the all-important recipes, the *Materials and Methods*. Next, collect your data and draft the *Results*. As your experiments end, formulate the outlines of a *Discussion*. Then write a working *Conclusion*. Now, go back and write the historical context, the *Introduction*. Only after all else has been written and tidied up, will you have sufficient perspective to write the *Title* and the *Abstract*.

Throughout the writing, your tools and techniques will be the same. You should use precise words and, whenever possible, numbers. You should write direct sentences that follow a straight line from point A to point B. In addition, you should fill all sections of the stereotyped skeleton of a standard scientific paper.

Writing a paper should be an active part of your research. If you wait until your studies are finished before you begin to write, you will miss a powerful tool. Research is iterative—you do, you assess, and you redo, and writing a paper is a way for you to continually make the reassessments necessary for critical and perceptive research.

Your manuscript can even be a blueprint for your experiments. The empty skeleton of a scientific paper poses a set of research questions, and, as you fill in the skeleton, you automatically carry out an orderly analysis of your data and observations. Moreover, by setting new data into the draft of your paper, you can maintain perspective. You will filter out the shine of newness, as your results—even unusual results—are put into the context of your existing data and your full research plan.

As a scientist, you must write, and, as an experimentalist, writing while you work strengthens your research. Writing a paper can be an integral part of observational science.

Scientific Papers Used as Examples

In the text of this book, I rebuild and improve a paper that I wrote in 1985, entitled "Intensifier for Bodian Staining of Tissue Sections and Cell Cultures." I use this paper because it is brief, simple, and well known to me.

Just as a picture is worth a thousand words, an actual example of a well-written scientific paragraph is worth a dozen descriptions of one. To illustrate the craft of scientific writing, I have included excerpts from scientific papers far better than my own.

The excerpts are from articles across the range of scientific studies. For the most part, these papers are lean, logical, and cleanly written. They are examples of good science writing and they have been recommended to me by the editors of the journals in which they appeared. In the text, I refer to the papers by author(s) and date. Here are the full bibliographic citations:

- Abercrombie M, Heaysman JEM. 1954. Observations on the social behaviour of cells. II. "Monolayering" of fibroblasts. Exp Cell Res 6: 293–306.
- Augspurger NR, Scherer CS, Garrow TA, Baker DH. 2005. Dietary s-methylmethionine, a component of foods, has choline-sparing activity in chickens. J Nutr 135: 1712–1717
- Berg D, Siefker C, Becker G. 2001. Echogenicity of the substantia nigra in Parkinson's disease and its relation to clinical findings. J Neurol 248: 684–689.
- Bohm A, Piribauer M, Wimazal F, Geissler W, Gisslinger H, Knobl P, Jager U, Fonatsch C, Kyrle PA, Valent P, Lechner K, Sperr WR. 2005. High dose intermittent ARA-C (HiDAC) for consolidation of patients with de novo AML: a single center experience. Leukemia Res 29: 609–615.
- Borgens RB, Bohnert D, Duerstock B, Spomar D, Lee RC. 2004. Tri-block copolymer produces recovery from spinal cord injury. J Neurosci Res 76: 141–154.
- Fastovsky DE, Sheehan P. 2005. The extinction of the dinosaurs in North America. GSA Today 15: 4–10.
- Gapski R, Barr JL, Sarment DP, Layher MG, Socransky SS, Giannobile WV. 2004. Effect of systemic matrix metalloproteinase inhibition on periodontal wound repair: a proof of concept trial. J Periodontol 75: 441–452.
- Glaunsinger B, Ganem D. 2004. Highly selective escape from KSHV-mediated host mRNA shutoff and its implications for viral pathogenesis. J Exp Med 200: 391–398.
- Haseler LJ, Arcinue E, Danielsen, ER, Bluml S, Ross D. 1997. Evidence from Proton Magnetic Resonance Spectroscopy for a Metabolic Cascade of Neuronal Damage in Shaken Baby Syndrome. Pediatrics 99: 4–14.
- Jacobson C-O. 1959. The localization of the presumptive cerebral regions in the neural plate of the axolotl larva. J Embryol Exp Morph 7: 1–21.
- Kiekkas P, Poulopoulou M, Papahatzi A, Panagiotis S. 2005. Is postanesthesia care unit length of stay increased in hypothermic patients? AORN J 81:379–382, 385–392.
- Milner B, Taylor L, Sperry RW. 1968. Lateralized suppression of dichotically presented digits after commissural section in man. Science 161: 184–186.
- Paul DR, McSpadden SK. 1976. Diffusional release of a solute from a polymer matrix. J Membrane Sci 1: 33–48.
- Perez JF, Sanderson MJ. 2005. The frequency of calcium oscillations induced by 5-HT, ACH, and KCl determine the contraction of smooth muscle cells of intrapulmonary bronchioles. J Gen Physiol 125: 535–553.

- Readinger ED, Mohney SE. 2005. Environmental sensitivity of Au diodes on n-AlGaN. J Electronic Mater 34: 375–381.
- Richards TW, Lembert ME. 1914. The atomic weight of lead of radioactive origin. J Am Chem Soc 36: 1329–1344.
- Rosenbaum DA. 2005. The Cinderella of psychology. The neglect of motor control in the science of mental life and behavior. Am Psychologist 60: 308–317.
- Rutherford E. 1919. Collisions of alpha particles with light atoms. IV. An anomalous effect in nitrogen. Lond Edinb Dubl Phil Mag J Sci 37: 581.
- Singer M, Weckesser EC, Geraudie J, Maier CE, Singer J. 1987. Open fingertip healing and replacement after distal amputation in Rhesus monkey with comparison to limb regeneration in lower vertebrates. Anat Embryol 177: 29–36.
- Speidel CC. 1932. Studies of living nerves. I. The movements of individual sheath cells and nerve sprouts correlated with the process of myelin-sheath formation in amphibian larvae. J Exp Zool 61: 279–317.
- Sugimori M, Lang EJ, Silver RB, Llinas R. 1994. High-resolution measurement of the time course of calcium-concentration microdomains at squid presynaptic terminals. Biol Bull 187: 300–303.
- Sundar G, Widom B. 1987. Interfacial tensions on approach to a tricritical point. J Phys Chem 91: 4802–4809.
- Williams CM. 1961. The juvenile hormone. II. Its role in the endocrine control of molting, pupation, and adult development in the Cecropia silkworm. Biol Bull 121: 572–585.

Part I
TOOLS AND TECHNIQUES

Chapter 1

THE STANDARDS OF A SCIENTIFIC PAPER

1.   A STEREOTYPED FORMAT

Research papers are the repositories of scientific observations plus the recipes used to make those observations.

Scientific papers have a stereotyped format:

- *Abstract*
- *Introduction*
- *Materials and Methods*
- *Results*
- *Discussion*
- *Conclusion*
- *References*

The exact section headings sometimes vary, but most scientific papers look pretty much the same from the outside. There are no novel constructions or inventive twists of the narrative. Instead, the framework is unchanging so that the content can be studied without distraction. The predictable form of a scientific paper, with its standard set of sections arranged in a stereotyped order, ensures that a reader knows what to expect and where to find specific types of information.

2.   PRECISE LANGUAGE

Within this stereotyped format, the language of a scientific paper aims to be clean, clear, and unemotional.

Much of the color of our everyday language derives from ill-defined, emotionally charged, ear-tickling images conjured up by sensuous words such as 'slovenly,' 'sibilant,' and 'sneaky.' Science, however, avoids colorful words.

The essential characteristic of scientific writing is clarity. Slippery words and vague phrases are confusing, and there is no place for ambiguity, arcane language, or froth in the archives of scientific records. In science, descriptions must be precise, recipes must be complete, data must be exact, logic must be transparent, and conclusions must be cleanly stated.

M.J. Katz, *From Research to Manuscript,*
© Springer Science+Business Media B.V. 2009

3.   A SINGLE, CLEAR DIRECTION

Beyond a stereotyped format and transparent language, a scientific paper also needs clarity of direction. Your entire paper should point inexorably toward its *Conclusion*.

Therefore, as you write, point the way for your reader, and remove tangents and digressions. Keep a single theme at the fore. For example, if your *Conclusion* is about temperature, then temperature should be ever-present in your paper. 'Temperature' should be in the *Title*. The *Introduction* should tell how your predecessors wrote about temperature. The *Materials and Methods* section should detail the instruments that you used and the operations that you performed involving temperature. The *Results* section should include data about temperature, and the *Discussion* section should connect your data to the existing scientific literature about temperature.

4.   REVIEWED AND MADE AVAILABLE TO OTHERS

Finally, a scientific paper should be accessible to others. Scientific journals are the traditional mechanisms for reviewing, disseminating, and preserving scientific papers, so submit your paper to a peer-reviewed journal. Having your paper reviewed by experts ensures that it can be understood and used by a broad scientific community. Then, having your paper preserved in a public forum ensures that the scientific community will have the opportunity to use it.

Chapter 2

SCIENTIFIC WORDS, SENTENCES, AND PARAGRAPHS

1.   SCIENTIFIC TEXT NEEDS EXACTNESS AND CLARITY

1.1.   Write with Precision

In science, your goal is to write a paper that is easy to understand. The art of scientific writing is not in the subtle underlying message conveyed by your prose. Instead, scientific prose is judged by how well it defines the details of the observations that you have made. In a short story, the reader might marvel at the "sensual writing, with hints of the mysteries of space and time." In a scientific paper, however, your prose style should disappear, and the reader should marvel at the realistic, explicit, and cleanly etched picture that you have painted.

Scientific papers have a stereotyped format so that there are no distractions from their contents. Likewise, scientific prose should be formulaic and plain. Here, the medium is not the message, the message is the message. Therefore, when you write a research paper, make your message precise and keep the medium unobtrusive.

To write precisely is to write without adornment. It can be an effort to recognize fluff and imprecision in your own writing, so train yourself to catch and to remove vagaries, emotion, indirectness, and redundancy. (For examples of the simplification of wordy phrases, see **Appendix B** below.)

It helps to remember that your goal is to speak plainly, i.e., to write clean straightforward sentences without hedging. Say what you mean directly. For example:

- "It may therefore not be unexpected that ..." should be "These results suggest ..."
- "An effort was made to ..." should be "We tried to ..."
- "The sorbitol probably acts to increase ..." should be "The sorbitol probably increases ..."
- "This gene is of significant interest for understanding commonalities in the evolutionary history of the microorganisms A and B" is clearer, simpler, and more informative when you tell exactly what you have in mind, such as "A single mutation in this gene of microorganism A has brought about its new use in microorganism B"

M.J. Katz, *From Research to Manuscript,*
© Springer Science+Business Media B.V. 2009

- "It is our considered opinion that other authorities may have misstated the relative import of such particulate concatenations in the soluble phase of the paradigm" should be written with specifics, such as "In their 1994 paper, Drs. Williams and Wilkins say that the drug's failures are due entirely to the clumping of suspended drug particles. In contrast, we propose that the viscosity of the solvent causes 40–50% of the failures."

1.1.1.   Use Numbers

Numbers have just the right properties for scientific writing: numbers are precise, objective, unambiguous, and without emotional undertones. Moreover, numbers can be used to describe many things in the real world; for example, in a variety of ways, numbers can represent shapes and sizes:

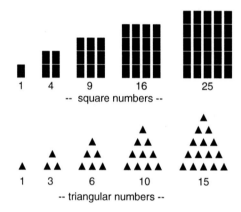

Because quantifiable adjectives are ideal descriptors in science, try to redefine all your adjectives as numbers. 'Tall' should be defined numerically, for example, 'greater than 2 m' or 'greater than 7 km.' Likewise, 'heavy' should be 'greater than 10 kg' or 'greater than 100 kg' or, perhaps, 'greater than 10^5 kg.' If you use 'brief,' tell us whether it means less than a minute, less than a second, or less than a millisecond.

Even the inherently subjective adjective 'painful' should be set as a number on a scale quantifying *how* painful, as is done in most hospitals:

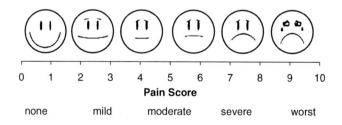

1.1.2. Use Objective Words

Of course, you cannot write with numbers alone. When quantifiable words are not available, you should use as precise and objective a vocabulary as possible.

Whether any particular sentence is precise and objective, depends on the reader's ability to define all its components. For example, "The needle vibrated continuously" is appropriate in a scientific paper if the reader is told which needle, what type of vibration, and over what time period it vibrated continuously. In science, the rule is, *define all your words*.

Beyond this rule, a few writing habits will help to ensure good scientific text. One of these habits is to weed out or replace vague and subjective terms; for instance, remove:

- Expressions with no clear limits, such as
  - o a lot, fairly, long term, quite, really, short term, slightly, somewhat, sort of, very
- Words of personal judgment, such as
  - o assuredly, beautiful, certainly, disappointing, disturbing, exquisite, fortuitous, hopefully, inconvenient, intriguing, luckily, miraculously, nice, obviously, of course, regrettable, remarkable, sadly, surely, unfortunately

- Words that are only fillers, such as
  - o alright, basically, in a sense, indeed, in effect, in fact, in terms of, it goes without saying, one of the things, with regard to

- Casual colorful catchwords and phrases, such as
  - o agree to disagree, bottom line, brute force, cutting edge, easier said than done, fell through the cracks, few and far between, food for thought, leaps and bounds, no nonsense, okay, quibble, seat of the pants, sketchy, snafu, tad, tidbit, tip of the iceberg

1.2. Scientific Use of Tenses

Good scientific prose uses a precise vocabulary. Scientific prose also uses verb tenses in a standardized way. When discussing research, the present tense indicates general knowledge and general principles, while the past tense indicates results of experiments.

1.2.1. Present Tense Is for Generalities

Use the present tense for general knowledge statements, widely accepted statements, and statements for which you could cite textbook references; for example:

- "Black-eyed Susan (*Rudbeckia hirta*), a member of the Aster family, is a plant native to North America."
- "Hexoses formed by digestion in the intestinal tract are absorbed through the gut wall and reach the various tissues through the blood circulation."

- "The term 'nuclide' indicates a species of atom having a specified number of protons and neutrons in its nucleus."
- "On a protein-rich diet, the amount of methylhistidine in the urine increases."

1.2.2. Past Tense Is for Specific Observations

Your results—the particular observations that you made during a research study—are bits of history, so use the past tense when you report your experimental results. For example:

- "In photographs of Guatemalan tarantulas, we found that the number of dorsal stripes ranged from six to nine."
- "During his war-time expedition to Guatemala, Rawski (1943) reported finding tarantulas with 9 stripes."
- "Eighteen percent of the patients in our study developed a mild rash."
- "The diodes were compared at regular time points during the next 75 h."

2. THE PARAGRAPH IS THE UNIT OF EXPOSITION

2.1. Each Paragraph Makes One Point

In a research paper, each paragraph should contain one main idea, and the space between paragraphs should be like taking a mental breath. Picture the text as, *Idea #1,* breathe, *Idea #2,* breathe, …

Most people absorb ideas in small chunks, and scientific paragraphs are those small absorbable chunks. You can assess the absorbability of a paragraph simply by counting its sentences. The ideal size for a paragraph is 3–4 sentences, and five sentences are about the upper limit. If you find that you have written six or more sentences without allowing for a mental breath, then go back and break your writing into smaller chunks.

Consider this paragraph about insulin.

- 'To keep all the cells in the body coordinated and working toward the same metabolic goals, the body uses hormones. Hormones are chemicals that are carried throughout the bloodstream, giving the same message to all the cells they meet. For sugar metabolism, the hormone messenger is *insulin*. Insulin is a protein that is made in the beta cells, which are clustered inside the pancreas. When the level of glucose in the blood becomes too high, the beta cells secrete insulin molecules into the bloodstream; thus, after a meal, the pancreas puts a large dose of insulin into the blood. The message that insulin then transmits throughout the body is "it's time to absorb, use, and store glucose."'

This paragraph contains six sentences, and its length alone should send you back to your writing desk. Reading the paragraph, you can find two major ideas. First, there are sentences about hormones in general. Second, there are sentences about one specific hormone, insulin. To emphasize each of these ideas, we should break the paragraph in two: one paragraph concerning hormones in general and the other concerning the nature and the effects of insulin:

- 'The body uses hormones to coordinate the metabolism of its many far-flung cells. A hormone is a chemical that is carried in the bloodstream and that gives a message to the cells it contacts. For sugar metabolism, insulin is one of the hormone messengers, and its message is "take up, use, and store glucose."
- 'Insulin is a protein that is made in beta cells, which are clustered inside the pancreas. When the level of glucose in the blood becomes too high, the beta cells secrete extra insulin molecules into the bloodstream. After a meal, for instance, the pancreas secretes a large dose of insulin into the blood.'

In a literary work, where the ebb and flow of words conveys a subconscious emotional message, a page of short paragraphs can be choppy and disruptive. However, a research paper has a different goal. Scientific writing must present a clear unemotional experience. Here, the methodical form, *Idea #1*, breathe, *Idea #2*, breathe ..., is an effective way to write.

2.2. Inside a Scientific Paragraph

2.2.1. The Lead Sentence

A typical scientific paragraph begins by stating its point, so the lead sentence should tell us the focus of the paragraph. In the two-paragraph example above, the first lead sentence, "The body uses hormones to coordinate the metabolism of its many far-flung cells," tells us that the first paragraph is about hormones as long-distance messengers. The second lead sentence, "Insulin is a protein that is made in beta cells, which are clustered inside the pancreas," tells us that the second paragraph is about a specific hormone, insulin.

2.2.2. The Subsequent Sentences

The remaining 2–3 sentences in each paragraph expand on the focal point that was identified in the lead sentence. Inside the paragraph, the sentences may:

- Give examples of the focal point.
- Give more details about the focal point.
- Remind readers that the focal point is a member of a more general class of similar things.
- Highlight an implication of the focal point.

In our example above, the first lead sentence tells us that the focal point of the paragraph is:

- HORMONES = LONG-DISTANCE MESSENGERS

The second sentence gives details of both sides of this equation:
- HORMONE = CHEMICAL
- HORMONAL MESSENGERS TRAVEL VIA THE BLOODSTREAM
    Finally, the third sentence gives specific examples:
- INSULIN = HORMONE
- INSULIN'S MESSAGE = "TAKE UP, USE, AND STORE GLUCOSE"

2.2.3.   Internal Flow

A scientist should be able to read your paragraphs without pausing. To give your writing this flow, each sentence of a paragraph should set the stage for the following sentence. Each internal sentence should be an extension of its predecessor. This can be done by making the subject or object from sentence number one the subject or object of sentence number two. By sharing its predecessor's subject or object, the second sentence continues the discussion and connects new ideas to those that have been established previously.

For example, in the first paragraph of the example above, 'hormone' is an object in sentence number one, and it is then used as the subject of sentence number two:
- "The body uses *hormones* to coordinate the metabolism of its many far-flung cells. A *hormone* is a chemical that is carried in the bloodstream and that gives a message to cells it contacts."

Likewise, 'hormone' and 'message' are a subject and an object in sentence number two, and 'message' is used as a subject and 'hormone messenger' is used as an object in sentence number three:
- 'A *hormone* is a chemical that is carried in the bloodstream and that gives a *message* to cells it contacts. For sugar metabolism, one of the *hormone messengers* is insulin, and its *message* is "take up, use, and store glucose." '

2.3.   Connect Succeeding Paragraphs

In the same fashion, you can smooth the travel between paragraphs by making the lead sentence of each paragraph refer to the previous paragraph. The flow between paragraphs is most natural if the subject of the lead sentence is a subject or an object in the last sentence of the preceding paragraph. In our example above, 'insulin' makes the bridge between the two paragraphs:

- 'The body uses hormones to coordinate the metabolism of its many far-flung cells. A hormone is a chemical that is carried in the bloodstream and that gives a message to cells it contacts. For sugar metabolism, *insulin* is a hormone messenger, and its message is "take up, use, and store glucose."
- '*Insulin* is a protein that is made in beta cells, which are clustered inside the pancreas. When the level of glucose in the blood becomes too high, the beta cells secrete extra insulin molecules into the bloodstream. After a meal, for instance, the pancreas puts a large dose of insulin into the blood.'

From sentence to sentence and from paragraph to paragraph, the flow of your argument should be linear:

- *Point A* implies *point B, point B* implies *point C, point C* implies ... symbolically:

$$A \rightarrow B, \ B \rightarrow C, \ C \rightarrow \ ...$$

Linear logic is the easiest for a reader to follow, so build your paper of paragraphs using simple linear logic. In the end, when they are strung together, these paragraphs should present a one-dimensional, step-by-step explanation of your experimental results.

Chapter 3

WRITING SCIENTIFIC TEXT

1.   BEGIN TO WRITE WHILE YOU EXPERIMENT

When you first sit down to write, you will not have a clean line of reasoning that explains your data. In the beginning, you may still be uncertain as to which are the most useful conclusions that can be argued from your observations. You may not even have identified the parts your data that are the strongest, the clearest, or the most thoroughly documented. With these or other issues still vague, it would be hard to write your paper linearly, beginning with the first paragraph of your *Introduction* and then adding paragraph after paragraph until you finally finish writing the last sentence of your *Conclusion*.

A great deal of intellectual work must be done before a tightly reasoned research paper can be completed. Rather than do all this mental work before you begin writing, however, you can discover the logic while you write. The process of writing a research paper can be exploratory, and it can even be a part of the research project.

Paper writing is an effective way to do the intellectual part of your research. As you write, you will organize your data, you will formulate explanations, and you will uncover connections between your results and the results of other scientists. Writing is a way to build the logical structure of and the scientific context for your experiments.

2.   START BROADLY, WORK ON THE DETAILS LATER

In your paper, you would like to show how to fit your data into a pattern that is revealing, satisfying, and perhaps a bit surprising. It takes time and trial-and-error to piece such a pattern together. You must wrestle with your data and your ideas, and if you are discovering your paper as you write, many drafts will come and ago before the form of your manuscript becomes stable and solid.

As you are creating the structure of your paper, it is not worthwhile to worry about the polish of your writing. When you begin work on your manuscript, put your polishing tools aside and step back from the details. Instead, look at your embryonic paper from a distance, as if through the wrong end of a telescope, so that you can see the broad sweep of your research. Then, start your work by blocking out thick chunks of ideas and arranging these chunks in a simple linear order.

With a string of blocks of ideas, your next step is to sculpt these rough chunks. Whittle away excesses and irrelevancies. Rearrange ideas, looking for simple patterns and natural connections. Carve and remodel, then step back and reassess. As you progress, shape finer and finer details, and find and highlight smaller and smaller interconnections. Meanwhile, be satisfied working with rough, imperfect sentences. Only at the very end, when the manuscript has settled into an organized linear narrative, should you polish the language.

As you take this global-to-local approach, work on one layer at a time, and do the actual work by breaking your writing into separate tasks. During one work session, collect piles of raw material—lists of ideas, notes, and facts. In another session, add logical connections by attaching elements of the lists together into statements. At a later session, introduce an additional level of logical organization by assembling the statements into rough paragraphs. Only the final sessions should be devoted to finding the precise wording that will make your paper crisp and readable.

When you begin writing, you will not have a clear vision of your paper, but this should not scare you. Without knowing the final shape of the text, you can dive fearlessly into the writing, because, by using a global-to-local strategy, the logical structure of your paper will emerge on its own.

3.   A MAGNIFIED VIEW OF THE WRITING PROCESS

As an example of global-to-local writing, the following sections rewrite the first paragraph of an old paper, "Intensifier for Bodian Staining of Tissue Sections and Cell Cultures" (Katz MJ, Watson LF, 1985, Stain Technol 60: 81–87). This paper reports a chemical intensification technique for a commonly used silver stain. The intensifier is useful because it improves the staining of embryonic axons[1] and of axons growing in tissue culture. Later in the book, I will again use parts of this paper as examples.

To expose the full writing process, I will try to describe every small step. Undoubtedly, you already take many of these steps unthinkingly as you write. However, things that are done unthinkingly can be more easily improved when they are made visible, so take a moment to look at this type of scientific writing in its most elementary steps.

3.1.   Use the Skeletal Outlines

Begin writing your paper one section at a time. Each section of a scientific paper has a stereotyped internal structure, a skeletal outline, and these skeletons are

---

[1] An *axon* is the long, thread-like extension of a nerve cell. Axons carry electrical impulses from one cell to the next. Each axon is microscopically thin, and when hundreds or thousands of axons are bundled together, the whole cable is called a nerve.

described in **Part II**, below. When writing the text of a section, start with an empty outline of its skeleton.

The skeleton of the *Introduction* section of a scientific paper is:

A. Background
1. Currently Accepted General Statements
2. Available Supporting Data
B. Gap
C. Your Plan of Attack

An *Introduction* begins by restating a general and well-accepted idea. From this known information, the section then leads readers to the particular unknown area, the scientific gap, that the paper plans to explore. The *Introduction* is a specialized historical essay, so the *Currently Accepted General Statements* subsection, which begins the *Introduction*, typically looks into the past.

> My paper was to be about staining axons, and I decided to divide the initial retrospective subsection into two topics, *General* and *Specific*. The beginning of my skeletal outline became:
>
> A. Background
> 1. Currently Accepted General Statements
>   a. General History of Axon Stains
>   b. Specific History of the Bodian Stain

3.2. Pile in Ideas

Now, take your outline and fill the empty spaces under each heading. List all the related ideas that come to mind. Don't worry about completeness or logic, and don't bother to write sentences.

Continue brainstorming and jotting down notes for the entire outline of the section that you are writing. Write all the ideas and facts that come into your mind, and don't stop until each heading is followed by at least three words or phrases.

> For the topic entitled "General History of Axon Stains," I tried to think of words and phrases about the classic work that has been done on axon staining. My initial list was:
>
> A. Background
> 1. Currently Accepted General Statements
>   a. General History of Axon Stains
>   • Reproducible
>   • State of the art is molecule-specific
>   • Organelle highlighting
>   • Signal amplification
>   • Individual cell stains a big leap
>   • Began with silver stains of Golgi used by Cajal

3.3.   Collect Information from Outside Resources

Next, go to your references—your books, articles, and notes. If you are work-ing on a part of your manuscript that is built largely from outside information, such as the *Introduction* or the *Discussion*, you will use books, articles, and databases. If you are working on a section built largely from your experiments, such as the *Materials and Methods* or the *Results*, you will be using your research records.

Search each reference for relevant information, and add these facts (with a note about their sources) under the appropriate headings of your outline.

> After doing this, my outline looked like:
>
> A. Background
>   1. Currently Accepted General Statements
>   a. General History of Axon Stains
>   • Reproducible
>   • State of the art is molecule-specific
>   • Organelle highlighting
>   • Signal amplification
>   • Individual cell stains a big leap
>   • Began with silver stains of Golgi and Cajal
> • Ref A – cresyl violet for neuron chromatin
>   neurons have strong affinity for weak silver solutions
>   Santiago Ramon y Cajal, 1890–1911 (Spain), summarized in "Degene-
>   ration and Regeneration of the Nervous System" 1928
>   Stephen W Ranson, 1914 (USA)
> • Ref B – David Bodian 1936 used silver protein and metallic copper, produced
>   clean stains of nerve cells, nuclei, axons, dendrites
> • Ref C – Camillo Golgi, Italian, late nineteenth century, pretreatment with
>   potassium dichromate, followed by silver nitrate, stains only fraction
>   of neurons and neuroglia and blood vessels
>   Golgi method gives 3-D view, good for cell architecture
>   Silver stains = best views of individual cells until electron microscopy
> • Ref D – Golgi, 1880, first used photographic processing techniques of Daguerre,
>   1839, to stain neural tissue

3.4.   Form Rough Sentences

Now complete the ideas. Take the words or phrases in your lists and replace them with full sentences, adding any information that is necessary to make a precise and informative statement. To write a complete and understandable sentence often requires some careful thinking, and you may need to consult your references again to fill your sentences with accurate details.

Continue writing full sentences for the entire outline of the section on which you are working.

For my paper, I wrote:

A. Background
  1. Currently Accepted General Statements
    a. General History of Axon Stains
      • Useful cell stains must give reproducible results.
      • A good stain will be specific for components of the feature of interest, not the entire feature itself. For cells, a good stain highlights specific parts of the cell, the membranes, or the various organelles. For membranes or organelles, a good stain highlights specific component molecules.
      • A good stain of tiny items gives a signal that is strong or that can be easily amplified.
      • For neuroanatomy, a big leap in understanding the architecture of the nervous system was the ability to stain an entire cell with all of its fine processes.
      • The study of individually stained nerve cells began at the end of the nineteenth century with the Italian histologist Camillo Golgi and the Spanish histologist and father of neurohistology Santiago Ramon y Cajal.
      • Cresyl violet proved a good stain for neuron cell bodies, highlighting the cell's chromatin.
      • Neurons have a strong affinity for the silver in weak salt solutions.
      • Between 1890 and 1911, Cajal meticulously documented the cellular architecture of the nervous system using silver stains. Much of his work is summarized in the English tome, "Degeneration and Regeneration of the Nervous System," published in 1928.
      • In the United States, Stephen W. Ranson began a series of silver studies on neural histology in 1914.
      • In the United States, in 1936, David Bodian introduced a simple and reliable silver stain for axons using solutions of silver protein with metallic copper. His stain produced clean staining of the nerve cell, its axon, and dendrites.

(continued)

(continued)

> • The use of silver stains for neurons was introduced in the late nineteenth century by the Italian histologist, Camillo Golgi. His technique pre-treated the fixed tissues with potassium dichromate and followed with a solution of silver nitrate. The Golgi technique was idiosyncratic, staining only a fraction of the neurons, neuroglia, and neural blood vessels. However, a stained cell usually revealed the full three-dimensional cell architecture.
> • Until the invention of electron microscopy, silver stains gave the best views of the three-dimensional structure of individual nerve cells (Ref C. Santini, 1975; Parent, 1996).
> • Golgi introduced his technique in 1880 and based it on Daguerre's 1839 procedures for processing silver-based photographs.

3.5.   Arrange the Sentences into Themes

You now have a list of complete statements. Your next task is to organize the statements into paragraphs.

In your finished paper, each paragraph will make a single point. The first step toward building focused paragraphs is to collect statements that concern a common subject or theme. Therefore, group related sentences and give each group a *Temporary Theme Label*, a *TTL*.

> In my rough draft, I divided my list of sentences into three themes:
>
> A. Background
>   1. Currently Accepted General Statements
>   a. General History of Axon Stains
>
> *TTL #1 General Features of a Good Cell Stain*
> • Useful cell stains must give reproducible results.
> • A good stain will be specific for components of the feature of interest, not the entire feature itself.
> • For cells, a good stain highlights specific parts of the cell, the membranes, or the various organelles.
> • For membranes or organelles, a good stain highlights specific component molecules.
> • A good stain of tiny items gives a signal that is strong or that can be easily amplified.
>
> *TTL #2 Neuron Stains*
> • Cresyl violet proved a good stain for neuron cell bodies, highlighting the cell's chromatin.
> • Neurons have a strong affinity for the silver in weak silver salt solutions.

(continued)

(continued)

> - For neuroanatomy, a big leap in understanding the architecture of the nervous system was the ability to stain an entire cell with all its fine processes.
> - Until the invention of electron microscopy, silver stains gave the best views of the three-dimensional structure of individual nerve cells (Santini, 1975; Parent, 1996).
>
> *TTL #3 The History of Silver Stains*
> - The use of silver stains for neurons was introduced in the late nineteenth century by the Italian histologist Camillo Golgi.
> - His technique pre-treated the fixed tissues with potassium dichromate and followed with a solution of silver nitrate.
> - The Golgi technique was idiosyncratic, staining only a fraction of the neurons, neuroglia, and neural blood vessels. However, a stained cell usually revealed its full three-dimensional cell architecture.
> - The study of individually stained nerve cells began at the end of the nineteenth century with the Italian histologist Camillo Golgi and the Spanish histologist and father of neurohistology Santiago Ramon y Cajal.
> - Golgi introduced his technique in 1880 and based it on Daguerre's 1839 procedures for processing silver-based photographs.
> - Between 1890 and 1911, Santiago Ramon y Cajal meticulously documented the cellular architecture of the nervous system using silver stains.
> - Much of his work is summarized in the English tome, "Degeneration and Regeneration of the Nervous System," first published in 1928.
> - In the United States, Stephen W. Ranson began a series of silver studies on neural histology in 1914.
> - In the United States, in 1936, David Bodian introduced a simple and reliable silver stain for axons using solutions of silver protein with metallic copper.
> - His stain produced clean staining of the nerve cell, its axon, and dendrites.

3.6.   Make Your Themed Lists into Rough Paragraphs

Now, take each themed group, and turn it into a rough paragraph. The typical scientific paragraph starts with a summary sentence, and the succeeding sentences expand the summary, step-by-step, so begin building paragraphs by writing the summary sentences.

The *Temporary Theme Label* was a first attempt at grouping your statements according to common topics. As you now review each themed group of statements, look again for the elemental common denominator. Throw away the original *Temporary Theme Label*, and read the group of statements anew, asking, "What is the best summary of this particular set of statements?" Then, write that summary as a simple sentence. This new summary sentence will be the *Lead Sentence (LS)* of your paragraph. Follow the *Lead Sentence* with the remaining sentences in an order that feels logical.

For my paper, I turned my three groups of sentences into these three rough paragraphs:

A. Background
  1. Currently Accepted General Statements
    a. General History of Axon Stains

LS #1 An ideal cell stain is detailed, reproducible, and strong.

An ideal cell stain is detailed, reproducible, and strong. A detailed stain reveals the internal structure of the object of interest. For the cell level, a stain should highlight the components of cells, such as the membranes and the various organelles. For the subcellular level, such as the organelles, a stain should highlight components such as molecules and molecular complexes. A reproducible stain gives the same results in different researchers' hands. A strong stain gives a signal that is easily detected macroscopically or that can be easily amplified.

LS #2 The architecture of individual nerve cells determines their function.

The architecture of individual nerve cells determines their function. For neuroanatomy, a big leap in understanding the architecture of the nervous system was the ability to stain an entire cell with all of its fine processes. Neuron cell bodies could be well defined by cresyl violet, which highlights chromatin. The extent and shape of the neuron's cell processes, however, are not seen with this stain. The entire neuron with its arborizing cell processes has a strong affinity for silver in weak solutions of silver salts. Silver staining, using variants of photographic developing techniques, gave the best views of the three-dimensional structure of individual nerve cells before the development of electron microscopic histology in the 1950s (Santini, 1975; Parent, 1996).

LS #3 The use of silver stains for neurons was introduced in the late nineteenth century by the Italian histologist Camillo Golgi.

The use of silver stains for neurons was introduced in the late nineteenth century by the Italian histologist Camillo Golgi. Golgi introduced his technique in 1880 and based it on Louis Daguerre's 1839 procedures for processing silver-based photographs. Golgi's specific technique pretreated the fixed tissues with potassium dichromate and followed with a solution of silver nitrate. The Golgi technique was idiosyncratic, staining only a fraction of the neurons, neuroglia, and neural blood vessels. However, a stained cell usually revealed the three-dimensional cell architecture. Beginning in 1890, the Spanish histologist Santiago Ramon y Cajal used silver stains to meticulously document the cellular architecture of the nervous system. Much of his work is summarized in English in "Degeneration and Regeneration of the Nervous System," published in 1928. In the United States, Stephen W. Ranson began a series of silver studies on neural histology in 1914. In the United States, in 1936, David Bodian introduced a simple and reliable silver stain for axons using solutions of silver protein with metallic copper. His stain produced clean staining of the nerve cell and its axon and dendrites.

3.7.　Take a Break and Clear Your Mind

It is time to take a break from writing.

As you wrestle to clarify partially formed ideas, the elements of your paper take on emotional tinges that come from your struggles. Gaps and problematic passages in your draft get blacker and bleaker and begin to feel like irritants. On the other hand, new insights, clean connections, ingenious ideas, and seemingly perfect bits of fact take on a golden shine and feel disproportionately important.

Time is a great balancer. As the days pass, the temporary emotional highlights in your work will fade, and the importance of various ideas will regain a more realistic proportion. Therefore, take many breaks—go away, turn your mind elsewhere, and let time tone down the vivid colors.

3.8.　Put Together One Paragraph for Each Topic

After a rest, pick up your draft again, with the goal of working through the entire outline of the section under construction, topic by topic.

Each topic now contains a set of rough paragraphs. Pick a topic and consolidate its paragraphs:

- First, decide which paragraph most directly addresses the main issue, and put this paragraph under the topic's title.
- Second, among your goals are directness and brevity, therefore if any of the remaining paragraphs deal with issues peripheral to the main point of the paper, toss them out.
- Third, try to merge the remaining paragraphs into the first paragraph. If that is too awkward, try to consolidate all the sentences into no more than two paragraphs, even if, at this stage, those paragraphs are long, cumbersome, or difficult to read.

In my example, I had three rough paragraphs for the first topic, *General History of Axon Stains*. This topic was to be the very beginning of my *Introduction*, describing the early work on the silver staining of cells. Only one of my three rough paragraphs was historical, so I chose this to be the main paragraph. I then took key sentences from the other two rough paragraphs and merged them with the main paragraph. Finally, I tossed out the leftover sentences.

The resulting paragraph was:

A. Background
   1. Currently Accepted General Statements
     a. General History of Axon Stains
The use of silver stains for neurons was introduced in the late nineteenth century by the Italian histologist Camillo Golgi. Golgi introduced his technique in

(continued)

(continued)

1880 and based it on Daguerre's 1839 procedures for processing silver-based photographs. Golgi's specific technique pre-treated the fixed tissues with potassium
dichromate and followed with a solution of silver nitrate. He found that the entire
neuron with its arborizing cell processes has a strong affinity for silver in weak
solutions of silver salts. The Golgi technique was idiosyncratic, staining only a
fraction of the neurons, neuroglia, and neural blood vessels. However, a stained
cell usually revealed the three-dimensional cell architecture. Beginning in 1890,
Santiago Ramon y Cajal used silver stains to meticulously document the cellular
architecture of the nervous system. The architecture of individual nerve cells—
specifically, the extent and shape of the neuron's cell processes—determines
their function. Much of his work is summarized in English in "Degeneration and
Regeneration of the Nervous System," published in 1928. In the United States,
Stephen W. Ranson began a series of studies on neural histo-logy in 1914. In the
United States in 1936, David Bodian introduced a simple and reliable silver stain
for axons using solutions of silver protein with metallic copper. His stain produced clean staining of the nerve cell nuclei, axons, and dendrites. Silver staining
using variants of photographic developing techniques gave the best views of the
three-dimensional structure of individual nerve cells before the development of
electron microscopic histology in the 1950s (Santini, 1975; Parent, 1996).

At the moment, your paragraphs will be wordy, long, and awkward, but they will
be fine for this stage in the writing process. Don't worry about the language, just plow
ahead, forming these rough foundation paragraphs for the entire outline of the section.

3.9. Shape a Working Draft

3.9.1. List Simple Sentences

To clean and tighten the large, rough paragraphs, you must build them afresh:
it is time to go back to lists.

Take each paragraph and list its sentences in order. If a sentence is complex,
break it into two or three consecutive simple sentences.

Make each sentence spare and declarative, and make them transitive statements
of the form: *Jane ate ginger cookies.* Use few and well-chosen adjectives, and try
to avoid adverbs.

In my draft, I turned my rough paragraph into this list of simpler sentences:

A. Background
   1. Currently Accepted General Statements
     a. General History of Axon Stains

(continued)

(continued)

> - Silver staining of neurons began in the late 1800s.
> - Camillo Golgi was an Italian histologist.
> - In 1880, he reported staining neural tissue with silver.
> - In 1839, Louis Daguerre had devised silver-based photography.
> - Golgi modified Daguerre's photographic processing techniques.
> - Golgi pretreated fixed tissues with potassium dichromate.
> - Then he soaked the tissue in silver nitrate.
> - The entire neuron has a strong affinity for silver salts.
> - Golgi's stain showed a neuron in three dimensions.
> - This delineated the full arborization of dendrites and axon.
> - The Golgi technique was idiosyncratic.
> - It stained cleanly only a small number of a tissue sample's neurons.
> - Santiago Ramon y Cajal was a Spanish histologist.
> - Beginning in 1890, Cajal studied silver-stains of neural tissues.
> - He mapped the cellular architecture of a wide variety of nervous systems.
> - The architecture of a neuron determines its function.
> - Degeneration and Regeneration of the Nervous System (1928) summarizes Cajal's work.
> - Stephen W. Ranson was an American histologist.
> - Beginning in 1914, he used silver stains to map neuroanatomy.
> - David Bodian was an American histologist.
> - In 1936, he reported a new silver stain for neurons.
> - Bodian's technique used solutions of silver protein with metallic copper.
> - His technique was simpler and more reliable than Golgi's was.
> - It stained the majority of axons in a tissue sample.
> - Silver stains give detailed views of the three-dimensional structure of individual nerve cells.
> - Until the development of electron microscopic histology, neurohistology depended on silver stains (Santini, 1975; Parent, 1996).

3.9.2. Remove Nonessentials

A well-written scientific paper is crisp and to-the-point. This is a good time to look critically at your text and to remove extraneous sentences.

Check each sentence against the point of the paragraph. Toss out any sentence that:

- Is tangential, with details unnecessary for a clear presentation.
- Has the same basic content as other sentences.
- Contains only non-scientific color or details of human interest.

From my list, I removed

Tangents
• Stephen W. Ranson was an American histologist.
• Beginning in 1914, he used silver stains to map neuroanatomy.

Unnecessary details
• Golgi pretreated fixed tissues with potassium dichromate.
• Then he soaked the tissue in silver nitrate.
• The architecture of a neuron determines its function.
• Bodian's technique used solutions of silver protein with metallic copper.

Duplicated ideas
• Silver staining of cells is a variant of photographic developing techniques.
• Silver stains give detailed views of the three-dimensional structure of individual nerve cells.

Nonscientific color and details of human interest
• Camillo Golgi was an Italian histologist.
• Santiago Ramon y Cajal was a Spanish histologist.
• David Bodian was an American histologist.

3.9.3.   Arrange Your Ideas in a Natural Sequence

Now, take each pared-down list and reorder the remaining sentences so that every sentence follows logically from the preceding sentence. If one of the sentences introduces a new idea, consider it to be the beginning of a new list and separate it from its predecessor by an empty line.

After doing this, my reordered lists became:

A. Background
1. Currently Accepted General Statements
   a. General History of Axon Stains

[List #1]
In 1839, Louis Daguerre had devised silver-based photography.
The entire neuron has a strong affinity for silver salts.
Silver staining of neurons began in the late 1800s.
In 1880, Camillo Golgi reported staining neural tissue with silver.
Golgi modified Daguerre's photographic processing techniques.
Golgi's stain showed a neuron in three dimensions.
His stain delineated the full arborization of dendrites and axon.

(continued)

(continued)

> [List #2]
> Beginning in 1890, Santiago Ramon y Cajal studied silver-stained sections of neural tissues.
> He mapped the cellular architecture of a wide variety of nervous systems. Degeneration and Regeneration of the Nervous System (1928) summarizes Cajal's work.
>
> [List #3]
> The Golgi technique was idiosyncratic.
> It stained cleanly only a small number of a tissue sample's neurons.
> In 1936, David Bodian reported a new silver stain for neurons.
> Bodian's technique was simpler and more reliable than Golgi's was.
> It stained the majority of axons in a tissue sample.
>
> [List #4]
> Until the development of electron microscopic histology, neurohistology depended on silver stains (Santini, 1975; Parent, 1996).

3.9.4.   Reassemble Paragraphs

You will now have one or more lists of sentences, each representing a single idea. Look at each list, and decide whether its main idea is necessary for this particular subsection of the skeletal outline. If the idea is out of place, then find the part of the outline where it belongs, and move it there.

> In my example, List #3 goes beyond general history and introduces the specific stain—the Bodian stain—that is the central subject of my paper. Therefore, I moved this set of sentences to a later topic that introduces the Bodian stain.

Finally, take each list, and string its sentences together again to form a paragraph. As you do this, you may find that there are some lists containing only a single sentence. Find homes for these orphans in one of the fuller paragraphs.

> I transformed my remaining lists into one paragraph:
>
> A. Background
>   1. Currently Accepted General Statements
>   a. General History of Axon Stains
>   In 1839, Louis Daguerre had devised silver-based photography. The entire neuron has a strong affinity for silver salts. Silver staining of neurons began in the late 1800s. In 1880, Camillo Golgi reported staining neural tissue with silver.

(continued)

(continued)

> Golgi modified Daguerre's photographic processing techniques. Golgi's stain showed a neuron in three dimensions. His stain delineated the full arborization of dendrites and axon. Beginning in 1890, Santiago Ramon y Cajal studied silver-stained sections of neural tissues. He mapped the cellular architecture of a wide variety of nervous systems. Degeneration and Regeneration of the Nervous System (1928) summarizes Cajal's work. Until the development of electron microscopic histology, neurohistology depended on silver stains.

3.10. Smooth Transitions

At long last, you have a draft of a section of your paper. Your draft will be a set of paragraphs organized in the form of the section's stereotyped skeleton. Your last task is to make the draft clean, readable, and logically consistent.

Start from the beginning of the section, read the sentences to yourself, and listen with your inner ear. Fix awkward words or phrases. Smooth the transitions between sentences, fill in missing links between ideas, and remove repetitive words and phrases.

> My first editing led to:
>
> A. Background
>   1. Currently Accepted General Statements
>   a. General History of Axon Stains
>
> Silver staining of neurons began in the late 1800s, when Camillo Golgi reported staining neural tissue with silver. In 1839, Louis Daguerre had devised silver-based photography. Golgi found that the entire neuron has a strong affinity for silver salts. Golgi then modified Daguerre's photographic processing techniques. Golgi's stain showed neurons in three dimensions. The stain delineated the full arborization of dendrites and axons. Santiago Ramon y Cajal used Golgi's stain and mapped the cellular architecture of a wide variety of nervous systems. "Degeneration and Regeneration of the Nervous System" (1928) summarizes Cajal's work. Until the development of electron microscopic histology, neurohistology depended on silver stains (Santini, 1975; Parent, 1996).

You now have the makings of a manuscript. It is time for some distance and objectivity, so take another break from writing …

3.11.   Polishing

3.11.1.   Rework the Entire Draft

It is a new morning, and here you are, at your desk once more. Pick up the draft of your manuscript, and work through the whole thing, one paragraph at a time. For each paragraph, ask:

- Does it describe a single idea?
- Is it self-contained?
- Does it start with a summary statement?
- Do the following sentences explain, expand, and develop the initial summary statement?
- Are there extraneous comments?

Face each problem, and do your best to fix it.

With my rough manuscript in hand, I went to work, consolidating sentences, cutting unessential words, and simplifying the exposition, Eventually, I ended up with:

A. Background
   1. Currenttly Accepted General Statements
     a. General History of Axon Stains
Silver staining of neurons began in the late nineteenth century, when Camillo Golgi (1880) modified Louis Daguerre's (1839) photographic development techniques for histology. Golgi found that the entire nerve cell has a strong affinity for silver salts, and his new silver stain highlighted the full three-dimensional arborization of a neuron's dendrites and axon. With Golgi's stain, Santiago Ramon y Cajal (1928) then comprehensively mapped the cellular architecture of a wide variety of neurons systems. These silver stain studies were the basis of all neurohistology before the development of histologic electron microscopy in the 1950s.

> At this point, I was tired of rewriting, and I thought that my latest paragraph was not too bad. However, I knew from experience that there is never an end to the possible improvements. To create text that speaks simply and clearly, you must polish the draft over and over. Therefore, I got a cup of coffee and began yet another phase of polishing.

3.11.2. Fix Specific Types of Problems

A good technique for late-stage polishing is to work on your manuscript with blinders and a magnifying glass. Don't read for overall meaning. Don't pay attention to the global features. Instead, concentrate on sentences and words, and pick a single task each time you sit down at your desk. Every time you pick up your manuscript, imagine that someone else has given it to you to correct the details only, and look at the writing with a finicky editor's eye.

3.11.2.1. Cut, Trim, and Simplify

In one of your work sessions, look only at the length of the paragraphs. A common problem is putting too many ideas into one, long paragraph. Break up big paragraphs into two or three smaller ones, each of which is short and focused on a single point.

At another session, search for nonessential words and cut them out ruthlessly. For instance, if you find the sentence, "He thoroughly investigated many avenues of staining," then trim it to be, "He tried many stains." Adverbs tend to be expendable. "We carefully pipetted," can be simply "We pipetted," and "The data largely support the hypothesis," should be "The data support the hypothesis at a confidence level of ...". **Appendix B**, at the end of this book, suggests simplifications for a variety of unnecessarily lengthy phrasings.

3.11.2.2. Add Active Verbs

Use a work session to invigorate your verbs. The meaning of a sentence is clearer when its verb is specific and active.

For instance, the sentence:

- "Insulin secretion is controlled by the amount of glucose to which beta cells are exposed."

is muddy. The verb phrase 'is controlled' is intransitive, and this makes us wait until the end of the sentence to discover the actor, i.e., who is doing the controlling. In addition, 'control' is a minimally informative verb, and we will learn more if we are told the kind or the mechanism of the control. An active verb tied to a specific mechanism will give a more informative sentence, such as:

- "High concentrations of glucose stimulate beta cells to secrete insulin."

Whenever you find a generic intransitive verb, such as,

- "Nerve cells were studied in our lab."

try to rewrite the sentence with a more specific active verb and with details about the action. For instance, you might rewrite this sentence as,

- "Using a camera lucida, we drew the shapes of nerve cells from the intestinal walls of fixed murine tissues."

or

- "Using time lapse photography, we recorded the changing shapes of nerve cells in tissue culture."

or

- "We measured the sizes of mitochondria in the squid giant axon."

3.11.2.3. Use Precise Adjectives

When you use an adjective, it should be specific and informative, and one of your polishing sessions should be devoted to sharpening your adjectives.

First, try to find numbers to replace any adjectives that have a range of specific appearances. For instance, "Subject A was fat" should be, "Subject A was 5′ 6″ tall and weighed 150 kg" or "Subject A had a body mass index (BMI) of 30.0 kg/m²."

When you cannot use numbers, upgrade your adjectives from vague and generic to precise and specific. For instance, words such as 'good' or 'important' are empty, and a phrase such as 'a good cell stain' is uninformative. Replace bland adjectives with specifics; for example, instead of writing 'good' in "a good cell stain," tell us the particular kind of 'good' that you have in mind, such as, 'easily detected with light microscopy', 'cell-specific', 'high contrast', 'non-toxic', 'non-fading', 'highly-reproducible', or 'easily-applied'.

3.11.2.4. Make Sentences Flow

Devote a session to the flow of your sentences. Read each paragraph quickly to see how smoothly the sentences move. Flow problems show up as places that snag your attention and distract from the content. If you find yourself stopping or rereading a passage, then you have stumbled over a problem.

The following are some typical problems that break the flow of a paragraph.

> *Problem*: A sentence seems to pop up out of nowhere.
> *Solutions*:
> 1. Add a more detailed explanatory introduction at the beginning of the sentence.
> 2. Put the sentence elsewhere in the text, where it will follow logically from the preceding sentence.

(continued)

Problem: You stop reading to look back at earlier words or sentences.
Solutions:
1. There may be a pronoun (*it, they, he, she, who, which, that*) that refers to more than one preceding noun. In this case, rearrange the sentence so the reference is clear—for example, change

   • "The pencil lay on the table, and it rocked back and forth."
to
   • "The pencil lay on the table and rocked back and forth."

2. The connection between two ideas may be missing. In this case, add a connector. For example, in one draft of my staining manuscript, I had written:
"For his silver stain, Golgi modified Louis Daguerre's 1839 photographic development techniques."
   When I reread this sentence, I stopped and found myself thinking, "Wait a minute—what *is* the connection between silver stains and photography?" I fixed the problem by specifying the connection:

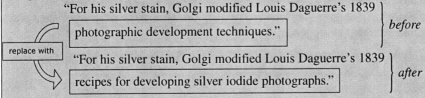

   "For his silver stain, Golgi modified Louis Daguerre's 1839 photographic development techniques."   *before*

   replace with

   "For his silver stain, Golgi modified Louis Daguerre's 1839 recipes for developing silver iodide photographs."   *after*

Problem: Ideas jump back and forth.
Solution: If a paragraph has a halting rhythm, see if rearranging the order of the ideas will smooth the flow. For example, when reading one of my drafts, I found myself being led from Camillo Golgi to Louis Daguerre and then bouncing back again to Golgi:
   • "Silver staining of neurons began in the late 19th century, when Camillo Golgi (1880) modified Louis Daguerre's 1839 recipes for developing silver iodide photographs. Golgi found that the entire nerve cell has a strong affinity for silver salts ..."
   I found that the narrative flowed more smoothly after I reordered the sentences to make the second, abrupt reference to Golgi disappear:
   • "Silver staining of neurons began in 19th century, when Camillo Golgi found that nerve cells have a strong affinity for silver salts. By 1880, he had succeeded in modifying Louis Daguerre's 1839 recipes for developing silver iodide photographs so that the silver would now stain neurons in fixed tissues."

3.11.3. Examples

As examples, here are repolished paragraphs from a variety of scientific articles.

• From a report on the apparent disintegration of nitrogen atoms by radioactive bombardment (Rutherford, 1919)

Original paragraph

"Since the anomalous effect was observed in air, but not in oxygen, or carbon dioxide, it must be due either to nitrogen or to one of the other gases present in atmospheric air. The latter possibility was excluded by comparing the effects produced in air and in chemically prepared nitrogen. The nitrogen was obtained by the well-known method of adding ammonium chloride to sodium nitrite, and stored over water. It was carefully dried before admission to the apparatus. With pure nitrogen, the number of long-range scintillations under similar conditions was greater than in air. As a result of careful experiments, the ratio was found to be 1.25, the value to be expected if the scintillations are due to nitrogen."

Polished paragraph

Because the anomalous effect is observed in air but not in purified oxygen or carbon dioxide, the effect must be due either to nitrogen or to one of the other gases present in atmospheric air. The non-nitrogenous atmospheric gases were excluded as causes by comparing the effects produced in air and in chemically prepared pure nitrogen. (See *Materials and Methods* for further details.) In pure nitrogen, the number of long-range scintillations was greater than in air. As a result of careful experiments, the ratio was found to be 1.25, the value to be expected if the scintillations are due to nitrogen alone.

- From a report consolidating evidence on the speed of extinction of North American dinosaurs (Fastovsky and Sheehan, 2005)

Original paragraph

"Like the dinosaur extinction, mammalian evolution in the early Tertiary of North America has been evaluated quantitatively. All agree that earliest Tertiary mammals underwent high rates of speciation leading to a steep increase in rates of diversification during the first 5 m.y. of the Tertiary (Fig. 4). Indeed, seventeen of the eighteen orders of extant placental mammals did not exist before the K-T boundary."

Polished paragraph

As has been done for the rate of dinosaur extinction, the rate of mammalian evolution has been quantified for the early Tertiary period in North America. These quantification studies suggest that mammals had high rates of speciation early in the Tertiary and that this resulted in the steep increase in the rates of mammalian diversification that have been found during the first 5 million years of the Tertiary (Fig. 4). As a result of this diversification, seventeen of the eighteen current orders of placental mammals arose after the K-T boundary.

- From a report on the effects of three natural agonists on the contraction of smooth muscle in lung airways (Perez and Sanderson, 2005)

Original paragraph

"Because a cholinergic pathway has been proposed as a mechanism by which 5-HT induces airway contraction in trachea and isolated lungs, we examined the effect of atropine on the 5-HT responses of lung slices to determine if any 5-HT effects occurred indirectly. Atropine ($1 \mu M$) had no effect on the airway contractile response (Fig. 7B) when added, either before or after exposure to $1 \mu M$ 5-HT. By contrast, $1 \mu M$ atropine totally abolished the airway contractile response induced by

1 µM ACH. Similarly, 1 µM atropine induced the full relaxation of an airway that was precontracted with ACH (Fig. 7B). These results indicate that 5-HT does not act via muscarinic receptors in the small airways of mice."

Polished paragraphs

Acetylcholine (ACH) will cause the smooth muscle in airways to contract, and it has been proposed that 5-HT causes airway contraction in trachea and isolated lung preparations by stimulating the release of acetylcholine. To explore this idea, we studied the effect of the anticholinergic drug atropine on the 5-HT responses of lung slices.

In our system, 1 µM atropine totally abolished the airway contractile response that is normally induced by 1 µM ACH. Similarly, 1 µM atropine fully relaxed an airway that was precontracted with ACH (Fig. 7B). In contrast, 1 µM atropine had no effect on the airway contractile response (Fig. 7B) when added, either before or after exposure to 1 µM 5-HT. These results indicate that 5-HT does not act via muscarinic cholinergic receptors in the small airways of mice.

- From a report examining the effect of matrix metalloproteinase inhibitors on healing after periodontal surgery (Gapski et al., 2004)

Original paragraph

"Although periodontitis is initiated by subgingival microbiota, it is generally accepted that mediators of connective tissue breakdown are generated to a large extent by the host's response to the pathogenic infection. In a susceptible host, microbial virulence factors trigger the release of host-derived enzymes such as proteases (e.g., matrix metalloproteinases [MMPs]) which can lead to periodontal tissue destruction. Collagenases, a subclass of the MMP family, are a group of enzymes capable of disrupting the triple helix of type I collagen—the primary structural component of the periodontium—under physiological conditions. Elevated levels of collagenases and other host-derived proteinases (e.g., cathepsins, elastase, and tryptases/trypsin-like proteinases) have been detected in inflamed gingiva, gingival crevicular fluid (GCF), and saliva of humans with periodontal disease."

Polished paragraph

The release of matrix metalloproteinases (MMPs), such as collagenases, by host tissues is part of the natural immune response to bacterial antigens and toxins. In humans with bacterial periodontal disease, inflamed gingiva, gingival crevicular fluid (GCF), and saliva contain elevated levels of collagenases and other proteinases (e.g., cathepsins, elastase, and tryptases). Collagenases are particularly destructive to type I collagen, which is the primary structural component of the periodontium. Thus, it is possible that at least some of the tissue destruction of periodontitis is caused directly by the host's response to the bacterial infection.

3.12.   When to Stop Writing

After a while, your text will become harder and harder to polish, and, eventually, you will hit the end of your ability to recognize more problems. It is time to stop and to let your text graduate.

> When I came to the end of my polishing ability, the first paragraph of my *Introduction* read
> A. Background
>   1. Currently Accepted General Statements
>   a. General History of Axon Stains
> Silver staining of neurons began in the nineteenth century, when Camillo Golgi found that nerve cells have a strong affinity for silver salts. By 1880, he had succeeded in modifying Louis Daguerre's 1839 recipes for developing silver iodide photographs so that the silver would now stain neurons in fixed tissues. Golgi's silver stain cleanly highlighted the full three-dimensional arborization of the axon and the dendrites of individual neurons. With Golgi's stain, Santiago Ramon y Cajal (1928) then mapped the cellular architecture of a wide variety of nervous systems. Cajal's comprehensive silver stain studies remain the foundation of modern neuroanatomy (Santini, 1975; Parent, 1996).

4.   ADVICE TO SPEAKERS OF OTHER LANGUAGES

Scientific logic is the same in all languages. If you are more comfortable using a language other than English, then write your paper in your own language first. After it is complete, translate it, or have someone else translate it, into English.

To make the final translation clearer, try to follow these suggestions when first writing your manuscript in your native language.

1. Words
   - Use simple verbs: write 'use' not 'employ.'
   - Turn adjectives into numbers: write '2' not 'several.'

2. Phrases
   - Don't use similes or metaphors, because they do not always translate properly. For example, write, "the mixture could not be poured" or "beads of the mixture stuck to the sides of the tube" not "the mixture was as thick as glue."

3. Sentences
   - Make each sentence short.
   - Put only one idea into each sentence.
   - Ignore the sound and the rhythm of the sentence in your native language, and don't try for smooth, flowing speech. Simple writing is easier to translate accurately than writing that sounds good to your ear.

4. Paragraphs
   - Make paragraphs short.
   - In each paragraph, arrange the sentences in direct logical order.

5. An English-Speaking Editor
   - After your paper has been translated, it is important to have it edited by a scientist who speaks English comfortably.

Chapter 4

PRESENTING NUMERICAL DATA

Scientific papers aim for objectivity, but this is a struggle, in part because words—even a scientist's words— come from a subjective vocabulary. One person's "hot" is another person's "warm," and one person's "pink" is another person's "red." Numbers, however, are different. Numbers are widely understood, and they have very similar meanings for most people, so numbers are well suited for objective writing.

Not only are numbers objective, they also have special non-linguistic powers. For instance, numbers can be ordered,

0 1 2 3 4 5 6 7 8 9 . . .

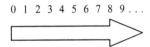

and ordering allows you to use the elemental comparisons *greater than*, *equal to*, and *less than* precisely and unambiguously.

For example, suppose that we have three sets of dots:

Do these three patterns have any natural sequence or order?

At the moment, we can order them in any way that we wish. However, if we specify a rule that assigns numerical values to the dot patterns, we can put them in a single, generally agreed upon order. For this, we might use the Braille rule, in which dot patterns are assigned numerical values as follows:

1   3   5   7   9

2   4   6   8   0

Now our three dot patterns have the numerical values 4, 6, and 8, and a natural order for them is:

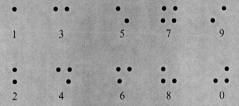

In addition to giving us the power of ordering, numerical statements can be embedded in an idealized abstract continuum, which is the basis for mathematical induction.

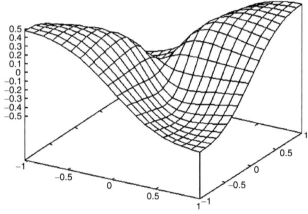

Mathematical induction allows you to slide along an endless continuum in mathematical space.

Mathematical induction is a mode of travel. It begins with a particular mathematical statement, and it then moves along a continuum by infinite, minuscule, automated steps. Using mathematical induction, you start at one place in the continuum and initiate a chain of iterative events, each set off by the previous event, like the falling of a line of dominoes.

Once started, the dominoes will fall along a preordained path stretching far into the distance. They will fall with an inevitability that assures you that, wherever they lead, it is to a place that is uninterruptedly connected to your starting point. With induction, you can travel great distances, and you can make reasoned numerical statements about places yet unvisited and things yet unseen.

Scientific writing is clearer when you take advantage of the objectivity and the far-reaching connectivity of numbers. Therefore, quantify your observations, and when you write, speak in numbers whenever possible.

1. TABLES

1.1. Organize Your Data

All real world phenomena are complex; therefore, when your experiments produce piles of data, you will always find variation and variety. Individual numbers have a clear meaning, but a pile of varied numbers can be harder to understand. It can take time and energy to see usable patterns in a large collection of numbers, so, when your data are numerous, you will help your reader by organizing your results before you present them in a report.

A good way to begin organizing your results is to put them into a table. A thoughtfully built table is the starting point for all scientific analyses, from qualitative discussions to sophisticated statistical and graphic presentations.

1.1.1. One-Dimensional Tables

A one-dimensional table is simply a list of your results. 'One dimension' means that each of your individual observations or data points is equivalent: they are all answers to the same question, "What happens when I do Experiment A?" The only distinctions between the results are their particular numerical values, and a natural way to organize a one-dimensional table of numbers is to list the values in numerical order.

1.1.2. Two-Dimensional Tables

A two dimensional table shows two features of each result. The two features might be the length and the weight of each output from the experiment (one full result would look like "1 cm, 2.5 g"), or the two features might be the volume of each output and the time when it was produced (one full result would look like "16 ml, 8 h after the start"). For a two dimensional table, each result is a pair of numbers, e.g. (1, 2.5) or (16, 8). A natural way to organize a two dimensional table of pairs of numbers is to list the pairs in numerical order according to the first number of the pair.

1.1.3.   Each Entry in Your Tables Should Be a Legitimate Experimental Variable

Each of the numbers in a single experimental result is called an experimental variable. One-dimensional tables present results reporting only one experimental variable, and two-dimensional tables present results reporting two experimental variables.

There is no theoretical limit to the number of variables that can be measured and reported in an experiment. There is, however, a restriction on what kinds of things can be legitimate scientific variables.

Consider this experiment. Suppose I want to see how many petals are produced by a new variety of black-eyed Susan (*Rudbeckia hirta*). In May, I plant 500 hybrid seeds. At the end of August, I count, by hand, the number of petals on each of the 500 flowers that have bloomed. To be certain that I count all the flowers, I label each flower with a unique number, and when I record my counts, I write each result as a pair of numbers—the flower's ID number and the flower's petal count. In the end, I have 500 pairs of numbers, and I organize them in the following two-dimensional table.

Characterizing a New Variety of Black-eyed Susan

Table variables
(a) Flower identification number
(b) Number of petals on that flower

Table

Flower id number	Number of petals
1	17
2	18
3	19
4	18
5	18
6	16
7	17
...	...

I have chosen to report two variables in this experiment. I could have chosen to report a thousand variables without violating any of the requirements of objective scientific study. (Although, with 500 sets of 1,000 numbers, I would have had to analyze quite a large pile [500,000] of numbers.)

On the other hand, the particular two variables that I chose to record pose a problem. Scientific experiments must be *repeatable*, and for the report of a scientific experiment to pass muster, its research procedures—its experimental recipes— must be explained clearly enough that a reader could follow the written instructions and get a similar set of results.

In my experiment, one of the reported variables was 'number of petals.' If you read the description of my procedure (given a few paragraphs earlier), and if you come to my greenhouse, then it seems likely that you can generate a list with the same petal counts that I found. The experimental variable 'number of petals' is something that:

• Another researcher could measure without much additional instruction
and
• On so doing, his results would probably match mine

However, this is not the case for my other variable, 'flower ID number'. My written explanation was,

• "To be certain that I count all the flowers, I label each flower with a unique number, and when I record my counts, I write each result as a pair of numbers— the flower's ID number and the flower's petal count."

I do not tell you how to decide which flower goes with which ID number, and the description does not give enough information for you to come to my greenhouse and assign my original numbers to the same flowers. Although you will probably have the same overall list of petal counts, it is unlikely that those petal counts will be matched to the same flower ID numbers that I used. Therefore, our two dimensional tables will probably differ.

I have recorded an experimental variable for which I gave an incomplete recipe, and part of my experiment cannot be repeated. If I write a scientific report about my work, I cannot report 'flower ID number' in the analysis of my data. Instead, my reportable results will be limited to a string of individual numbers, the 'number of petals on each flower,' and my *Results* can only contain a one-dimensional table.

This is an example of the principle that some variables that a scientist might consider recording are not acceptable experimental variables. The rule for choosing a legitimate experimental variable is straightforward:

• Legitimate experimental variables are the output of repeatable, explicitly described operations.

'Number of petals' is a legitimate variable because it is data discovered by following a research recipe that has been fully described in the protocol for the experiment. 'Flower ID number' is not a legitimate variable—at least, it is not legitimate until I provide a recipe that other researchers could use to generate the same ID numbers.

> What recipes would generate scientifically legitimate flower ID numbers? Some examples include:
> - Recipe a: flowers are numbered in the order that they bloom
> - Recipe b: flowers are numbered in the order of their planting dates
> - Recipe c: flowers are numbered in the order of their heights
> - Recipe d: flowers are numbered in the order of their diameters
> - Recipe e: flowers, which were grown evenly spaced along a North-South line, are numbered according to their GPS coordinates
>
> In these cases, 'flower ID number' is a legitimate experimental variable, because there is sufficient information for another researcher to reproduce my flower numbering system. However, 'flower id number' is legitimate only because it is a stand-in for a more direct and descriptive variable, such as 'date of blooming' or 'height'. Therefore, in these cases, it will be more informative if I skip the intermediary ID numbers and report the actual variable that I measured.

1.2.   Inside Tables Use Numerical Order

When you build your tables, take advantage of the intrinsic order of numbers. For one-dimensional tables, list the numbers numerically. If your data are, for example:

- (17) (18) (19) (18) (18) (16) (17)

list them instead as

- (16) (17) (17) (18) (18) (18) (19)

For some experiments, this list will be quite long. In an experiment with only a few results, you have room to report the entire numerically ordered list. On the other hand, in an experiment, such as my fictional black-eyed Susan project, with a great many results, the list of results cannot easily be packed into a concise research paper. Here, you should report a summary of the list.

For a one-dimensional table, a useful summary is a list of the frequency of occurrence of each of the different numbers. For example, if the results are:

- (16) (17) (17) (18) (18) (18) (19)

the summary table—the frequency distribution—is

Result	Number of occurrences
16	1
17	2
18	3
19	1

For the entire hypothetical black-eyed Susan experiment, the frequency distribution table might be:

Table 1

Number of petals	Number of flowers
15	6
16	54
17	114
18	221
19	91
20	14

Total = 500

Data
Counts of the number of petals on 500
hybrid flowers
Experimental Variable
The number of petals on each flower

Similarly, you can make a frequency distribution list from results with more than one variable. If I had designed my black-eyed Susan experiment to record two legitimate variables for each result, then I could report the frequency distribution in a matrix. For example, suppose that each result—i.e., each different flower—was characterized by both the height of the plant and the number of petals. Here, each result would look like '22 cm, 18 petals'. I would have 500 such results, and I could summarize them in a distribution matrix that might look like:

Table 2

– Number of plants with N petals –

Height (cm)	N = 15	N = 16	N = 17	N = 18	N = 19	N = 20
<11	5	1	0	0	1	0
11–15	0	34	26	27	0	0
16–20	1	18	68	167	75	6
21–25	0	1	20	26	13	1
>25	0	0	0	1	0	7

Total = 500

Data
Counts of the number of petals on 500 hybrid flowers
Experimental Variables
(a) The number of petals on each flower
(b) The height of each plant

To recap the main points about tables

- A table is always a good foundation for the organization and analysis of numerical results.
- Build your tables of legitimate experimental variables, i.e., numbers that come directly from recipes in the *Materials and Methods* section of your paper.
- Inside your tables, arrange the data in numerical order.

1.3.   Examples

The overall form of a data table is similar among scientific journals, but the formatting details can differ. Some journals use figure legends written in sentences or phrases, such as:

Distribution of Number of Petals

Number of petals	Number of flowers
15	6
16	54
17	114
18	221
19	91
20	14

Total = 500

Table 1. Our new hybrid of *Rudbeckia hirta* has yellow flowers that appear much like the wild variety. Of 500 hybrid flowers grown in a controlled temperature and humidity green-house, 44% had 18 petals. Eighty-five percent had 17, 18, or 19 petals. No flowers had less than 15 or more than 20 petals.

Other journals use internal footnotes as the legends for their tables, for example:

Table 1

Distribution of Number of Petals on Hybrid *Rudbeckia hirta*

NP*	NF†
15	6
16	54
17	114
18	221
19	91
20	14

Total = 500

*NP = number of petals/flower
†NF = number of flowers with NP petals

For footnotes inside tables, use these symbols in sequence:

* † ‡ § ‖ ¶ ** †† ‡‡

Further information about figure legends can be found under **Figure Legends** in **Chapter 5.2** below.

2.   STATISTICS

Using statistics properly is a skill. Even prepackaged statistical programs need to be used thoughtfully. No matter how much practice you have had with numerical analysis, never be shy about asking for advice from researchers with more experience. If you are fortunate enough to have access to professional statisticians, consult them from the very beginning of your experiments.

In the following sections, I introduce the vocabulary and basic tools that you will need for statistical analyses. To get beyond the basics, I recommend studying Freedman D, Pisani R, Purves R, 2007, *Statistics*, 4th ed. W.W. Norton, New York.

2.1.   Descriptive Statistics

When numerical data pour out of your experiment, it helps to reduce the volume to a few characteristic numbers. These characteristic numbers are *descriptive statistics*.

Descriptive statistics are more than just summaries of the numerical data points. Descriptive statistics characterize the whole pile of data—they view a data pile as a thing, an entity with its own particular size, shape, and texture.

A statistical analysis takes the data pile, re-orders it,

and offers numerical descriptions of the entire ordered set of data.

Certain descriptive statistics are especially helpful and frequently used, because they give you a direct intuitive feel for your data pile; for example:

Size

- The size is the total number of data points in the pile. Size is often represented by 'N.' The size of my pile of flower petal data, from above, is N = 500.

Range

- The range is the distance between the smallest and the largest data values. The range is the data pile's full width. In my flower petal data, the range is minimum = 15 petals to maximum = 20 petals, or range = 15–20 petals.

Middle

There are three commonly used middles.

- **Mean** The mean is the center of mass, the balancing point of the data. The mean is the average data value. For my flower petal example, mean = 17.8 petals.
- **Mode** The mode is the data value that occurs most often. The mode is the data pile's maximum height. In the flower petal example, mode = 12 petals.
- **Median** The median is a number that half of the data values are less than and half are greater than. The median divides the data pile in half. In the flower petal example, median = 12 petals. (If there is an odd number of data values, the median is a whole number. If there is an even number of data values, the median is the average of the two middle numbers, which can be a fraction.)

Spread

The spread is the compactness or dispersal of the data pile. If the numbers have little variation, the pile will be compact, whereas if the numbers are more heterogeneous and variable, the pile will be dispersed and spread out.

- **Standard Deviation** The standard deviation is a commonly used measure of the spread of a pile of numbers. A standard deviation is built from the deviations of each data point from the mean. A small standard deviation indicates that the pile

Standard Deviation
One particularly useful statistical diagram of a data pile shows the frequency of occurrence of each value laid out as a graph. These graphs are called frequency distributions or histograms. (See **Histograms** below.)
Here is a histogram showing data from a study of the birth weights of babies. The hatched bars indicate how many babies were born in each weight class during the study. Most of the newborns (113) were in the 3,133–3,393 g (7.0–7.3 lb) weight class—i.e., the *mode* is the 3,133–3,393 g interval.

(continued)

(continued)

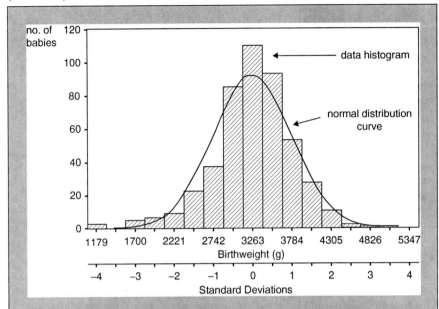

Drawn over the histogram is a continuous line showing how an idealized, bell-shaped *normal curve* (a normal distribution) might be fitted to the actual experimental data.

The standard deviation for the idealized curve is 521 g (1.1 lb). The bottom line shows the locations on the x-axis of 1, 2, 3, and 4 standard deviations away from the *mean*, which is 3,263 g. The properties of the standard deviation tell us that 68% of all the data values will be within −1 and +1 standard deviations from the mean of a normal distribution and that 95% of all data values will be within −2 and +2 standard deviations from the mean.

of numbers is compact, while a large standard deviation indicates that the pile of numbers is spread out.

All statistics programs, most spreadsheet programs, and many hand-held calculators can compute standard deviations.

- **Central 50%** The central 50% is a more intuitive measure of the spread of a pile of numbers. The central 50% shows the limits of the middle half of a data pile. When the range of the central 50% is narrow, the pile of numbers is compact, and when the central 50% has a wider range, the data pile is spread out.

Central 50%

The central 50% of a data pile lies between the 25th and 75th percentile values. To find these percentile values, line up your data points in numerical order. Find the data point that 25% of the data points are less than, and identify the value of that point as the lower limit of the central 50%. Then find the data point where only 25% of the data points are greater, and identify the value of this point as the upper limit of the central 50%.

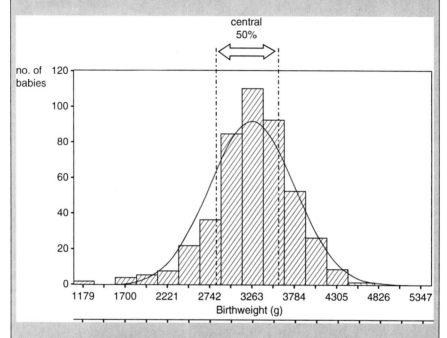

For the birth weight data, the central 50% of the weights is between 2,909 and 3,617 g (6.4–8.0 lb). The width (i.e., the range) of the central 50% is 708 g (1.6 lb). In this study, the average (mean) newborn weighed 3,263 g (7.2 lb) and half of all the newborns weighed between 2,909 g (6.4 lb) and 3,617 g (8.0 lb).

2.1.1.   Confirmatory and Exploratory Data Analysis

Some scientific projects are straightforward documentations. For these, you set up an observational method—a specific research protocol—and run the protocol without interference until you have collected sufficient data to properly document the protocol's results. Usually, you begin with a question that can be rephrased as a practicable recipe, and you simply follow that recipe. For instance, the question, "What is the average blood pressure of 65 year old male Caucasians in California in 2009?" might be rephrased as:

"What is the result of following the recipe:

Step (1) Identify a 65 year old male Caucasian in California

Step (2) Have him sit and rest for 5 minutes
Step (3) Take the blood pressure in his right arm while he is still sitting
Step (4) Repeat for 1,000 different males."

This documentation form of scientific work is sometimes called *confirmatory* data analysis, and it is the classic technique for proving or disproving a hypothesis. In confirmatory data analysis, you rephrase your hypothesis as a well-defined recipe, and you carry out the recipe with no changes. For instance, if you have hypothesized that the average blood pressure of 65-year-old male Caucasians in California is 135/85 mm Hg, you formulate a recipe for taking blood pressure measurements and you then carefully follow the recipe, without variations, until you have enough data to test the strength of your hypothesis.

Hypotheses can be tested by confirmatory data analysis, but testing a hypothesis is the second stage of the scientific method. The first stage is generating the hypothesis. Just as they can be tested by scientific research, hypotheses can also be generated by scientific research. This form of scientific work is sometimes called *exploratory* data analysis.

Many research projects are explorations. Typically, exploratory research is an iterative process, in which you stop occasionally to take stock of the situation, looking at where you have been and what you have seen and trying to discern emerging patterns in your data.

If your data are numbers, a useful way to study your accumulating data is by calculating descriptive statistics for your entire data set and for a variety of subsets at each reassessment. Different descriptive statistics represent different ways of summarizing your data set, and in your search for emerging patterns, you will cast your net most widely if you calculate many different descriptive statistics on many different subsets of your data.

2.1.2. Statistical Software for Exploratory Data Analysis

Exploratory statistical packages are wonderful tools for probing piles of numerical data. Software packages such as *JMP* (SAS) are visual and interactive, with easy-to-use graphical programs. If you have access to *JMP* or a similar program, use it to look at your data right side-up, upside-down, and from all other angles, searching for patterns.

During your exploratory analysis, you should step away from your preconceived ideas and look at your data with naive eyes. Ask yourself:

- Does the data set as a whole have a pattern?
- Do certain subsets of the data have patterns?
- Which data presentations make the patterns look simple?
- Which data presentations highlight surprises?

If you find something that catches your eye, write it down in the form of a specific hypothesis, such as,

- "The data appear to form a straight line."
- "The data appear to cluster around two central values."
- "The mean and the mode of the data points appear to be identical."

- "The data from time period #1 appear to increase logarithmically, while the data from time period #2 seem to form a straight line."
- "The data appear to be heterogeneous and patternless."

When you are exploring, don't stop with one or two hypotheses. As a well-known geologist, Thomas Chamberlin, pointed out, a good researcher is the parent of a large family of hypotheses (Chamberlin. 1890. The method of multiple working hypotheses. Reprinted 1965 *Science* 148: 754–759).

2.1.3. Descriptive Statistics Can Be Sufficient

"Exploratory data analysis," wrote the statistician John Tukey, "does not need probability, significance, or confidence [estimates]." In other words, when you are examining your data for patterns, descriptive statistics by themselves can be the appropriate numerical characterizations.

Even in your final scientific paper, when you are describing the clearest pattern that you see in your data, you can often rest on descriptive statements. A further step,

For example, suppose that your experiments generate two piles of numbers, and it looks to you like pile #2 has the same shape as pile #1. The most striking difference between these piles is that the numbers in pile #1 are clustered around the value 2.5, whereas the numbers in pile #2 are clustered around the value 4.6. For your hypothesis, you might write:

- "Hypothesis—Data set #2 has the same shape as data set #1, although data set #2 is shifted a fixed distance, 2.1 units, to the right in the plane."

In support of this hypothesis, you can offer numerical descriptions, i.e., descriptive statistics, of the data sets, such as:

Descriptive statistic	Data set #1	Data set #2
N	155	159
Mean	2.5	4.6
Median	2.5	4.6
Mode	2.5	4.6
Standard deviation	0.6	0.6
Width of central 50%	0.5	0.5

These descriptive statistics show specific ways in which the data sets are alike and specific ways in which they are different, and these quantified comparisons comprise a perfectly acceptable statistical analysis that you can use to give more precision to the narrative version of your hypothesis. No probabilities or significance values are needed in this presentation.

confirmatory data analysis—the hypothesis testing techniques of inferential statistics (see **Inferential Statistics and Hypothesis Testing**, below), which are often seen in research papers—can give you probability estimates (*p* values), significance limits, and confidence intervals. However, confirmatory data analyses require assumptions that make their conclusions at least one level more distant from the actual data. Descriptive statistics are closer to your data, they give a more direct numerical picture of the patterns that you see, and they tend to be the cleanest summaries of your results.

2.1.4. Include a Numerical Picture

Statistical summaries are the standard ways to describe piles of numbers. However, even the most hard-core statistician accedes to the special explanatory power of direct pictures. (See **Constructing Scientific Figures** in **Chapter 5**, below.) Humans excel at visual pattern recognition. With a glance, we can quickly see the order in complex or messy things, and just as automatically, our eye is drawn to any imperfection amidst neat and orderly things. We do these things far better than we manage the cumbersome processes of mathematical analysis. As John Tukey pointed out, "graph paper [exists] as a recognition that the picture-examining eye is the best finder we have of the wholly unanticipated."

For this reason, you cannot make a stronger case in support of a hypothesized data pattern than by demonstrating it visually, as in a graph.

For example, for the two data sets described in words and numbers a few paragraphs earlier, the addition of a picture gives your hypothesis a different kind of clarity and meaning.

Narrative description
"Hypothesis—Data set #2 has the same shape as data set #1, although data set #2 is shifted a fixed distance, 2.1 units, to the right in the plane."

Numerical description

Descriptive statistic	Data set #1	Data set #2
N	155	159
Mean	2.5	4.6
Median	2.5	4.6
Mode	2.5	4.6
Standard deviation	0.6	0.6
Width of central 50%	0.5	0.5

Visual description

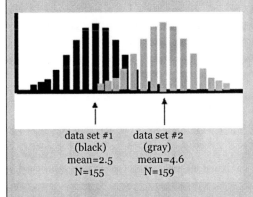

data set #1   data set #2
(black)   (gray)
mean=2.5   mean=4.6
N=155   N=159

In the *Discussion* sections of your research papers, you will often be comparing your data set to simple patterns, such as, lines, curves, or waves, or to other pre-existing data sets. These comparisons will be hypotheses, such as,

- "our data appear to be linearly distributed"
- "our data appear to differ from the data reported by Dr. Jones"
- "our experimental and control data appear to have the same mean"
- "our experimental and control data appear to have distinctly different means."

Together, the triad of a narrative description (a precisely written hypothesis), a numerical description (descriptive statistics), and a visual description (a graph), usually offers a strong explanation of your data.

Therefore, even in the exploratory stages of your data analyses, make a three-fold record of your impressions. Each time you formulate a tentative hypothesis, describe it in words, numbers, and pictures:

- Write a sentence describing the hypothesized pattern as precisely as possible.
- Calculate descriptive statistics that describe the pattern numerically.
- Draw a picture or graph of the data that shows the pattern.

As you finalize your research paper, you will choose one of your exploratory hypotheses to present as the best pattern that you have been able to see in your data. Your supporting arguments should include the threefold description from your research records.

2.2.   Inferential Statistics and Hypothesis Testing

For experiments that produce numerical data, you should search for mathematical reference patterns that appear to match your data. Mathematical reference patterns are patterned sets of numbers that come from simple, understandable formulas, such as $y = ax + b$, which is a formula for a line, $(x - a)^2 = 4p(y - b)$, which is a formula for a parabola, and $(x - a)^2 + (y - b)^2 = r^2$, which is a formula for a circle. When you find a mathematical reference pattern that resembles your data, write the hypothesis, "My data closely match the simple mathematical pattern P."

You can support such a hypothesis by presenting a clean narrative description, descriptive statistics, and a graph. However, the traditional way to characterize the *strength* of this kind of mathematical hypothesis has been by using inferential statistics, and for this, you must introduce additional assumptions.

When you have hypothesized that your data closely match a particular mathematical pattern P, inferential statistics begins by presuming that:

- A stochastic (probabilistic) process has generated the numbers you have captured in your experimental data set.
- This stochastic process produces numerical data in a pattern that matches the mathematical pattern P.

With these assumptions, inferential statistics allows you to calculate a number that ranks the confidence you can have in your hypothesis.

2.2.1. The Normal Distribution

The fundamental role that stochastic processes play in inferential statistics has prompted the addition of certain stochastic-friendly patterns to the list of simple mathematical reference patterns. The most prominent of these additions is the *normal distribution*.

The normal distribution is a curve with a smooth, symmetrical, bell shape.

Normal Distribution

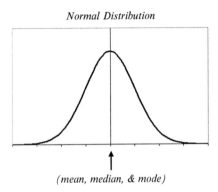

(mean, median, & mode)

In a normal distribution, the mean, the median, and the mode each have the same value, and the exact shape of any particular normal distribution can be summarized with just two numbers, its mean and its standard deviation.

The normal distribution is a mathematical abstraction. The flow of the distribution moves smoothly from value to value. The ends of the distribution extend infinitely in both directions from the mean. In the real world, no experimentally generated set of numbers can be arrayed so that their values grade with infinite smoothness along a continuum from point to point, and no set of data numbers tails off gently and infinitely toward its extreme values.

Real data do not perfectly match the pure normal distribution. Nevertheless, the mathematical properties of normal distributions make them tempting tools to statisticians. Moreover, large sets of real world data often resemble normal distributions; for example, a graph of the birth weights of babies has a shape that is much like a normal distribution:

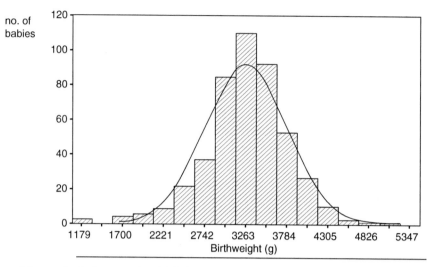

Many calculations made with inferential statistics are dependent on the assumption that your data set can be well approximated by a normal distribution. It is important to remember, however, that sets of real world data need not be normally distributed. For example, a histogram of the number of people dying at each age is not symmetric: instead, it is heavily skewed toward old age. Similarly, scores on college tests are usually concentrated asymmetrically in the upper third of the values, when calculated as percents.

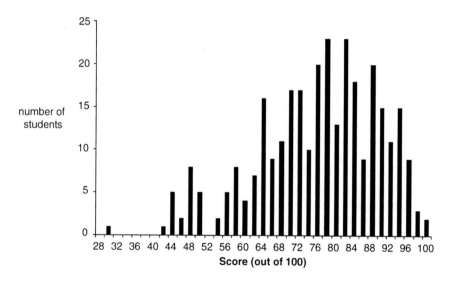

Although many real world data sets appear similar to normal distributions, other data sets more closely match different distributions that can also be used as the basis of statistical calculations; these distributions include:

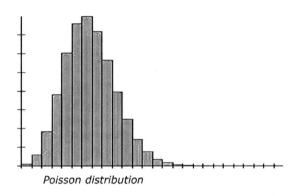

Poisson distribution

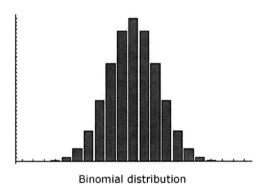

Binomial distribution

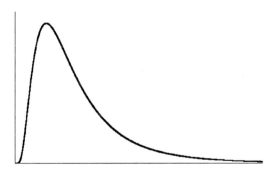

Lognormal distribution

For many statistical tools, you must know in advance which distribution best approximates your data. Therefore, before you begin using inferential statistics to calculate things like p values and confidence intervals, take an objective look at the shape of the distribution of your data. If you are not certain which distribution most closely matches your data, ask a statistician for help.

2.2.2.   Significance Tests

Inferential statistics includes a variety of significance tests that can be used to rank the strength of statements such as, "My data are normally distributed" and "My data are aligned along a straight line."

Typically, these rankings are given as p values, which are numbers on a scale of 0 to 1 indicating the probability that your experimental data is a subset of a specified mathematical formula or distribution. For instance, '$p < 0.05$' means that the probability is less than 5% that your data set is a part of the specified mathematical distribution. (Sometimes, p values are loosely rewritten as confidence levels—a $p < 0.05$ is a confidence level of 0.95 [or 95%] that your data set is *not* a part of the specified mathematical distribution.)

> By convention and when there are no other requirements in the experiment, two significance levels—$p < 0.05$ and $p < 0.01$—have been called "significant" in the texts of many science papers. These thresholds can be arbitrary, however, and they may not correspond to what we would consider to be meaningfully, functionally, or usefully significant.

Be sure to use the word 'significant' thoughtfully in your paper. If you decide to use inferential statistics, don't simply apply an arbitrary convention and say, "these data are significant" or even "these data are significantly different from [set *B*, a normal distribution, a line, a random distribution, …]." Minimally, you must follow the word 'significant' with the particular p that you have calculated. For example write, "These data are significantly different from [set *B*, a normal distribution, a line, a random distribution, …] ($p < 0.02$)" and then include details of your assumptions and your methods for calculating the p value.

2.2.2.1.   Parametric Tests
Significance tests that assume that the experimental data set matches a particular predefined distribution are called *parametric tests*. One of the commonly used parametric significance tests is the *t-test*. The t-test assumes that data piles have the shape of a normal distribution.[2]

---

[2] Comparable parametric tests are available for different mathematical distributions, such as the binomial distribution, the Poisson distribution, and the lognormal distribution.

A t-test compares two piles of numbers, each generated by a different experimental protocol. The t-test asks whether the means of the data piles are so different that it is very unlikely that the experimental protocols are equivalent. For instance, you can use a t-test to compare these two data sets:

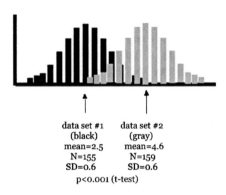

data set #1   data set #2
(black)   (gray)
mean=2.5   mean=4.6
N=155   N=159
SD=0.6   SD=0.6
p<0.001 (t-test)

Interpretation: The chances are less than 1 in 1,000 that data set #1 and data set #2 were generated by equivalent experimental protocols (assuming that these protocols produce data values that are normally distributed).

t-test formulas are found in statistics texts and statistics programs. They require three descriptive statistics for each pile of numbers: the *size*, the *mean*, and the *standard deviation*.

When presenting the results of a t-test, the full description would read:

- "A t-test found that the two data sets differ with a significance level of $p < 0.001$."

This full description is usually abbreviated as:

- "The difference was significant ($p < 0.001$, t-test)."

Remember to be precise. With this and many other significance tests, you are not testing whether two experimental protocols are the same. You are only testing whether you have shown that the two protocols may be different, a result which would be suggested by a small p value, such as $p < 0.0001$.

When your significance test gives a large p value, such as $p > 0.5$, it does not mean that there is no difference between the two experimental protocols. A large p value only means that your way of looking at things—your specific analytic technique—has not been able to detect a statistically strong difference between the data sets. Therefore, when your comparison gives a relatively large p value, i.e., $p > 0.5$, write, "Under these experimental conditions, a significant difference could not be detected ($p > 0.5$, t-test)." Don't write, "These data sets are statistically the same ($p > 0.5$, t-test)."

2.2.2.2.   Nonparametric Tests

Not all statistical tests assume that your data are normally distributed or, in fact, are distributed in any particular way. Some tests, such as the *chi-squared test*, are *nonparametric*. While they are based on additional underlying assumptions, nonparametric significance tests are distribution-free. Nonparametric tests that are analogous to t-tests include the *sign test*, the *Wilcoxon's matched-pairs signed-ranks test*, the *median test*, and the *Mann–Whitney U test*.

Nonparametric tests tend to be easy to use. A good example is the *Kolmogorov–Smirnov (K–S) test*, which is a simple non-parametric test for comparing two number piles. The *K–S test* has few requirements, and it can distinguish number piles that have different shapes, not just different means.

2.2.2.3.   Choosing a Significance Test

It is important not to let the statistics drive your analysis, so don't begin with a specific significance test in mind. When you have numerical data, you need a hypothesis before you can use inferential statistics, which are forms of confirmatory analysis.

First, look at your data and formulate a blunt hypothesis, a clean declarative statement with a single subject and a single object. Typical hypotheses are:

- My data lie along a straight line.
- My data are normally distributed.
- My data match a specified mathematical distribution.
- My data are distributed so heterogeneously that they are essentially random.
- My data match the distribution of a data set generated by Dr. Smith.
- My data generated using Recipe A match the distribution of another data set I generated using Recipe B.

With a hypothesis in hand, you can now look for a helpful significance test. You want a test that can use your particular data set and can evaluate your specific hypothesis.

Remember, when using inferential statistics, you will have to expand your hypothesis. Typically, you must add an assumption of this form:
- "I assume that the mechanism that generates each of my data points has a certain amount of imprecision. This imprecision means that the value of any one data point cannot be predicted exactly. However, the variability occurs randomly and causes the actual data points to occur in a normal distribution around a predictable mean."

Ensuring that the necessary assumption is scientifically credible requires more than a superficial understanding of statistics. This is a good time to talk with a statistician.

2.2.3. Software for Inferential Statistics

Calculating significance levels and other numerical evaluations of your hypotheses can be easy using a software package. However, inferential statistics always involves underlying models and assumptions. To understand the models, the assumptions, and the full meaning of your calculations, it is best to talk with a statistician. Therefore, I do not recommend using a software package of significance tests without experienced advice.

(In contrast, statistical packages with exploratory tools are wonderful aides. Software packages such as *JMP* [SAS] are visual and interactive, with easy-to-use graphical programs, and I recommend that you use exploratory software throughout your research.)

2.3. Use Statistics Thoughtfully

Statistics is carried out in the mathematical universe, where it operates on abstract, idealized images. The language from the mathematical universe is then carried over into our explanations of real world phenomena, and formulas inadvertently become interpreters of our data. With the mathematical images in mind, we find ourselves saying that data should look a certain way; for example, we may call some data points "good fits" and others "outliers."

The mathematization of real world phenomena pervades science. Various formulas from the mathematical universe are held to be idealized versions of physical processes that we see in the real world. For instance, the path of a thrown ball is often said to be a parabola, if there is no spin on the ball. But in the real world there *is* spin on the ball and there are also irregularities on its surface, inhomogeneities in its weight distribution, air resistance, wind, barometric changes, and the occasional sudden rainstorm. The precisely delineated processes of the mathematical universe that give rise to parabolas are only pale images of the complex, messy, interactive processes that generate data in the real world.

Your data set has been generated in the real world. It will rarely fit the idealized images that we dream of in the abstract mathematical universe. Don't let your explanations of real world data be driven by those mathematical idealizations that are of necessity embedded in all statistical tools.

Chapter 5

CONSTRUCTING SCIENTIFIC FIGURES

1.   BASIC GUIDELINES

Much of science is expressed through words and numbers. At the same time, humans are visual creatures, and inside our brains, pictures can be understood without words or numbers. For instance, you automatically absorb things from this picture of a plane flying over the grand canyon:

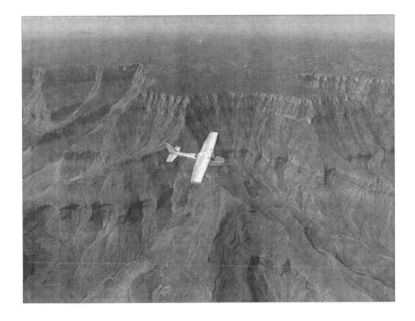

and this picture of Mount Everest:

that cannot easily be put into words or numbers.

In the human brain, pictures go directly to visual centers, where they register in the topology of the brain's cellular architecture. Pictures tell us things unavailable through words or numbers. For this reason, you can bring more and different information to your readers by including pictures.

A wide range of pictures—photographs, diagrams, drawings, graphs—can be figures in a scientific paper. Their content will determine their location in the text. Figures that portray techniques belong in the *Materials and Methods* section, figures of data belong in the *Results* section, and figures of synthetic concepts, abstractions, theories, and models belong in the *Discussion* section. As with tables, figures must be referenced and explained in the text, and you should number your figures consecutively in the order of their text citations.

Even within the confines of a formulaic scientific paper, you can be creative with your figures. The most important constraint on your ingenuity is that a scientific figure must be informative not entertaining. The substantive content of your figure—be it a diagram of a procedure, a photograph of your experimental subject, a graph of your data, or a drawing of your hypothesis—must be clear and in the forefront.

As to the form of the figure, there are a few general conventions that you can use as guides:

- Try to build your figures within an imaginary rectangle 1 unit high and 1.5 units wide.

- Make the information flow from left to right, and array numbers so their magnitudes increase from left to right.

- The frame of the figure should be unobtrusive. In graphs and similar figures, baselines (such as zero levels) and a scale should be indicated modestly at the bottom and at the left of the diagram.

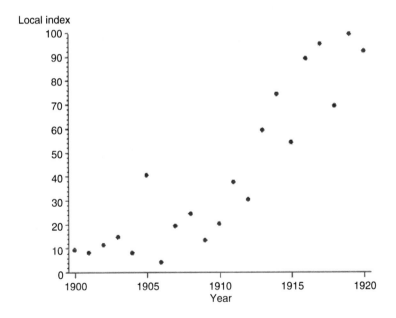

- Whenever possible, position words horizontally, not vertically, and make the lettering clear, unadorned, and well spaced.
- Number each figure, and give it a title and an explanatory legend. The legend should explain all the symbols and abbreviations and the essence of the figure's

content. Imagine that your figure will be used in a slide presentation, and include enough information so that the slide can be understood by itself.
• Overall, a figure should look simple, clean, and professionally produced.

2.   FIGURE LEGENDS

Each figure needs a legend, and, together, a figure and its legend should be self-explanatory, a miniature report in themselves.

Write figure legends in spare phrases—complete sentences are not necessary. The first phrase in your legend acts as the title, and it should summarize the figure. Succeeding phrases or sentences should give essential details. Definitions of symbols and abbreviations should follow the narrative explanation, and the legend should end with an indication of scale, if it is needed.

Example of a figure and its legend

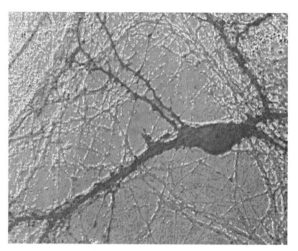

Figure 2. An embryonic neuron and neurites in culture. The fixed culture was stained using Bodian with 10 min post-staining intensification. (DIC microscopy, x 300)

3.   NUMERICAL FIGURES

3.1.   Graphs

When your data are numbers, graphs will show more visual detail than will tables. Tables order data in lines, whereas graphs organize data in a full two-dimensional plane, so try to use two-dimensional graphs when you can. (Two-dimensions are generally best—it can sometimes be difficult for people to see patterns in three-dimensional graphs.)

3.1.1. Graphing Software

You can make small charts and graphs using spreadsheet software, such as Microsoft's *Excel*. More versatile, dedicated software is also available; an excellent example is OriginLab's *Origin*.

3.1.2. Histograms

One simple type of graph is the histogram. A histogram, also called a frequency-distribution diagram, shows how many of your data points have each value. The range of data values is laid out along the x-axis, and the numbers of data points having each value are listed along the y-axis. (In a histogram, the x values [the classes or measurement intervals for the values] must be divided into equally sized intervals.) For example:

HISTOGRAM OF WEIGHT DISTRIBUTION OF ADULTS

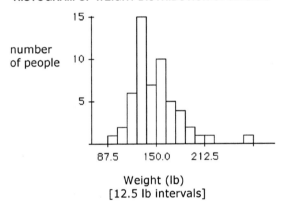

In **Chapter 4**, I presented a set of flower petal data as a table:

Distribution of Number of Petals

Number of petals	Number of flowers
15	6
16	54
17	114
18	221
19	91
20	14

total = 500

Table 1. Our new hybrid of *Rudbeckia hirta* has yellow flowers that appear much like the wild variety. Of 500 hybrid flowers grown in a controlled temperature and humidity greenhouse, 44% had 18 petals. Eighty-five percent had 17, 18, or 19 petals. No flowers had less than 15 or more than 20 petals.

The data can also be presented as a histogram, where each flower is a data point and the number of petals on the flower is the value of that data point.

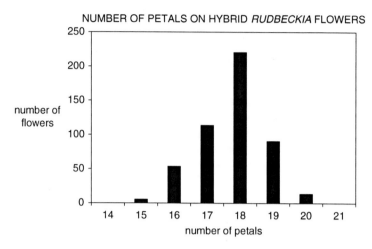

Graph 1. A histogram of the number of flowers that have particular numbers of petals. All flowers had between 15 and 20 petals (x-axis). Of the 500 flowers examined, 221 (44%) had 18 petals. Most flowers (83%) had either 17, 18, or 19 petals. The *mean* =17.8 petals, the *median* =18 petals, and the histogram is asymmetric.

By expanding the table into a graph, I have turned a numerical pattern into a spatial pattern. It takes some thinking to understand the numerical pattern in Table 1, but using visual logic, we can immediately comprehend the spatial pattern in Graph 1.

A histogram is simple because it portrays only a single aspect or variable of your data points. When your research project records two or more variables of each data point, you can make separate histograms of each of the individual variables. For example, if I had measured the height of the entire plant in addition to counting the number of petals on its flower, I could draw a second histogram showing the number of plants in each of the different height ranges. For example:

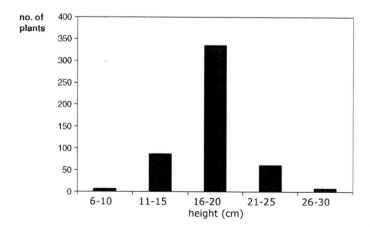

3.1.3.   Scatter Plots

These two histograms—petal number and plant height—present the data as if it were collected in two independent experiments. This is a useful way to view your data. When more than one key variable has been recorded from the same data set in the same experiment, it is a good idea to make an independent histogram of each variable so that you can examine the pattern of the occurrence of that variable alone.

In addition, when your experiments record two key variables for each observation, you can use another graph to look at the pattern of the co-occurrences of the variables. In a graph of co-occurrences, each observation is treated as a data point, and the position of the point in the plane is determined by the values of its two variables. The data points will now be scattered throughout the graph, so this graph is called a scatter plot or scatter diagram.

**A SCATTER PLOT OF THE CO-OCCURRENCE OF TWO VARIABLES
MEASURED IN THE SAME EXPERIMENT**

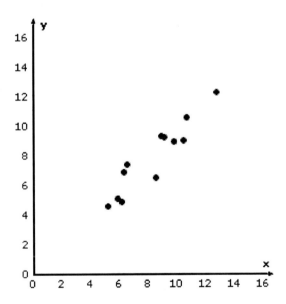

For example, a scatter plot of the co-occurrence of petal number and plant height, in my hybrid plant example, would have 500 points, one for each flower in the study:

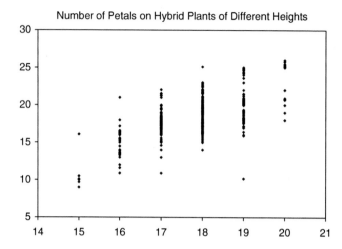

Graph 2. A scatter plot of the number of petals on flowers of plants with various heights. Five hundred hybrid *Rudbeckia* plants were measured. Taller plants tended to have flowers with more petals. The range of petal numbers was 15–20. The range of plant heights was 10.7–25.9 cm.

3.2.   Relationships Between Variables

A scatter plot challenges you to wonder whether the two key variables that you have graphed are correlated. Mathematically, 'correlated' means that changes in one of the variables are related to changes in the other variable.

The question hidden underneath the search for a correlation is, "Can the same mechanism be generating both variables?" For instance, suppose that I have measured:
- The number of petals on the flower of a plant
- The height of the plant

By searching for a correlation, I am wondering whether both values—i.e., petal number and plant height—are driven by the same biological cause. If I were to find that taller plants tend to have more petals, then I can begin to hunt for a common underlying cause. For example, perhaps *both* petal number and plant height have been affected by the amount of sunlight, water, or fertilizer or by the time since seed germination or by the proximity to electromagnetic fields.

Of course, it is always possible that I will not find a correlation. The only relation between the two variables may be that I chose to measure them each time I collected data. If I had decided to count the number of petals on a flower and simultaneously to record the Dow Jones Industrial Average at that moment, it is unlikely that I will have been collecting variables that are both influenced by the same causes, and I would be surprised to find a correlation.

3.2.1. Find a Pattern

When you have measured two variables for your data points, the scatter plot of co-occurrences of the variables is a good place to start looking for correlations. Begin the search with your eyes and your innate ability to see spatial patterns.

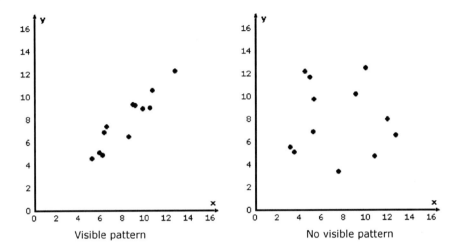

Look at your graph, and ask yourself these questions:

- Does it look like the points (or a subset of the points) in the scatter plot have some order? For example,
  - o Are the points clumped?
  - o Are the points more concentrated in certain areas?
  - o Do the points form a shape?
- Do the blank areas form a pattern?

And, if you think you see a pattern, try to put a name to it. For instance, does it look like a simple mathematical shape, such as, a line, a circle, a parabola, or a wave?

Then, before using any mathematical tools, put your impressions into words. Commit yourself by writing in your notes, "I see a pattern," or "I do not see a pattern," and if you see a pattern, write, "The pattern looks like _____" filling in the blank as unambiguously as possible, with phrases such as,

- A random dispersal of points
- A straight line
- A parabola
- A sine wave
- A circular patch of points
- A shapeless clump of points
- Dense points on the left grading to sparse points on the right

This sentence is your rough hypothesis.

3.2.2.　Write a Clear Hypothesis

With scatter plots, as with all the analytic parts of your research, finding and describing patterns is the way that you will make sense of your experimental data. The process is iterative. You say to yourself, "This group of data points seems to lie on a straight line," and you take that statement as a working hypothesis, to bolster or to discredit as you continue working.

When the nouns and the adjectives in your hypothesis are specific and precise, it will be easier to assess how much credence to give to this hypothesis. For a scatter plot, take your pattern statement and push it to be as precise as possible. For example, if your first pattern statement was:

- "The pattern looks like a straight line."

add as many details as you can. Perhaps you can write:

- "The pattern looks like a straight line, beginning at the data point (1, 13) and ending at the data point (7, 3). One data point, (3, 12), is farther from this line than any of the other 24 data points."

Or, if your first pattern statement was:

- "The pattern looks like one clump of points."

You might be able to write:

- "The patterns appears to be a roughly spherical cluster of data points, centered at approximately (50, 29). Specifically, all the data points are contained within a circle centered at (50, 29) with a radius of 18 graph units, and, to my eye, the data points appear fairly uniformly distributed within this circular area."

When the words in your hypothesis are well-defined, tests of your hypothesis will be more convincing.

Your aim is a threefold description. You now have a graph and a narrative description—two of three elements. Now, formulate a numerical summary of your data by calculating some descriptive statistics. Write down the size, the range, and the numerical middle (mean, median, and mode) of the points of your scatter plot. Then, give us a numerical summary of the dispersal of the points (standard deviation or central 50%). Finally, document any special aspects of the narrative description with actual numbers (e.g., "Ninety-five percent of the data points are greater than 15 and less than 175.").

3.2.3.　Publish the Full Triad

When your experiment generates a pile of numerical data, it is important to put all three descriptions into your *Results* section.
- Write a precise narrative description.
- Calculate descriptive statistics.
- Draw a histogram, scatter plot, or other graph.

If, in your Discussion, you propose that your data are arrayed in a pattern, you will point to the triad as both a description and a demonstration of that pattern.

In the descriptive triad, the narrative description is your hypothesis. For data sets composed of numerical values, you can sometimes use inferential statistics to add a significance level, i.e., a p value, to your hypothesis. Remember, when calculating significance values, you will have to make additional assumptions. Be sure to write these assumptions in your paper, so that your reader can assess their scientific credibility. (See **Inferential Statistics and Hypothesis Testing** in **Chapter 4**, above.)

3.3.   Aesthetics of Numerical Figures

Edward Tufte (Tufte, 2001, *The Visual Display of Quantitative Information*, 2nd ed., Graphics Press, Cheshire, CT) gives a superb survey and critique of the styles of pictures that have been used to present numerical information.

In his book, Tufte points out that figures should devote most of their ink to the data. In other words, most of the content of a drawing should be data, not framework, gridlines, decorations, or background.

Explanatory features, such as highlighting, arrows, and inside-the-graphic notes should be sparse. Artistic additions, such as filigrees to enhance data points or three-dimensional expansions of one-or two-dimensional objects, distract the reader from the actual information. Keep the drawing clean, airy, and junk-free, says Tufte. Let the eye take in the pure data, both in its broad sweep and in its detail. For example, this figure:

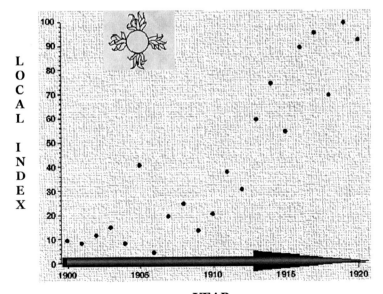

YEAR

should be replaced by this figure:

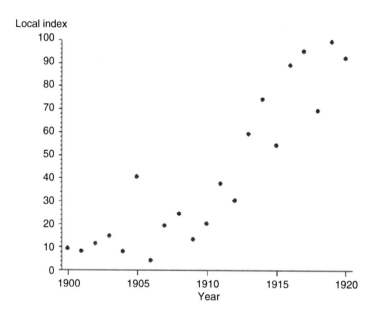

4.   PREPARATION FOR SUBMISSION TO A JOURNAL

Journals vary in their requirements for submitted figures, so wait until you have chosen a journal before you make the polished figure. Print journals will usually require photographs of each figure—either physical prints or photographic quality digital prints. Some print journals and all online journals want electronic files of figures in a particular format (e.g., JPEG) and at a sufficiently high resolution so that they can be inset directly into a website. Photographs can be easily polished and prepared for publication with any one of a number of software programs, such as Adobe's *Photoshop Elements*.

5.   SCIENTIFIC PATTERNS SHOULD BE
REPRODUCIBLE

Patterns in a data set can jump out at us when the data is presented visually, as in this graph:

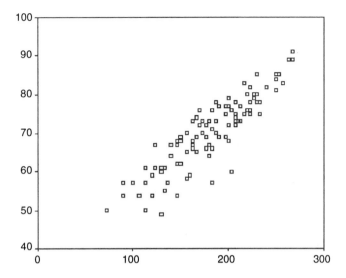

Nevertheless, as a device for revealing patterns in complex data sets, visual impressions are two-edged swords. Our brains will create images where no reproducible patterns exist, and we can see mirages—clumps, lines, curves, waves, shapes, faces—in most graphs that are thick with data, even in randomly-distributed data, such as this random scatter-plot:

This leads to a general caution about any pattern that you might discover. The universe is large, and its patterns are uncountable. Science, however, is concerned with only a small fraction of the patterns in the universe.

Science deals in those patterns that are reproducible, that can be checked and verified, that are a stable or repeatable feature of the systems under examination. Scientifically, it is useless for you to describe today's pattern of raindrops on the ocean. Raindrops splash in endless unpredictable patterns, and your descriptions will be drawn from Nature's bottomless well of unique, one-time patterns.

Science has no use for the fleeting and unique. A pattern that passes by only once is irreproducible and nonpredictive. We cannot build with it. It will not help us to manipulate or to control our environments.

Therefore, when you identify a pattern in a graph or in any other representation of your data, you must show your reader that the pattern is likely to be a persistent characteristic of your experimental system. In practical terms, you should either

- Demonstrate the same pattern when you repeat your experiments.
- Cite independent examples of the same pattern being associated with your experimental system.
- Use statistical techniques to argue that this particular pattern is unlikely to have occurred by chance.

Part II
WRITING A RESEARCH PAPER

Chapter 1

WRITING DURING RESEARCH

Research winds and backtracks. Experiments go in unanticipated directions, irregular and peculiar data pop up and threaten to derail what you thought were well-laid plans, and technical complications monopolize your attention. New questions materialize, your ideas change, you have second thoughts, and you doubt your hypotheses. Disorder constantly gnaws at the daily life of an experimental scientist.

Writing drafts of your scientific paper *while* you are experimenting helps to keep your day-to-day research orderly. Writing can keep your efforts and your ideas connected and clear, because:

- When you put your ideas into sentences, you have to face their logic (or lack of logic).
- When you sort your data into tables, you can see the holes that remain in your results.
- When you commit your recipes to the page, you are forced to record *all* the details.
- When you look at a skimpy or lopsided reference list writ large in black and white, you are embarrassed into doing more background research.

In addition, the skeleton of a developing scientific paper is an effective blueprint. It is a guide for pulling together a coherent story from the disparate activities that go into real world research. You will make more progress and operate more efficiently if you write while you work.

M.J. Katz, *From Research to Manuscript,*
© Springer Science + Business Media B.V. 2009

1.   KEEP A COMPUTERIZED NOTEBOOK

Your research notebooks should be computerized, and they should include at least two specific databases: a diary and a reference record.

1.1.   The Diary—Record Your Work Notes

A natural way to organize your work notes is by date. Whenever you do something in relation to your research project, take a moment to record your ideas and observations in a computerized diary. There are many commercial software programs for diaries. Pick one that is simple, that organizes information by date, and that allows you to insert a variety of objects, including text and pictures. A good freeware program is *EverNote* (EverNote Corp.), which can be downloaded from *http://www.evernote.com*

Two types of diary entries require some extra forethought: notes about your experimental techniques and records of your results.

1.1.1.   Experimental Techniques, the *Materials and Methods*

As you assemble your equipment, record the sources of your materials and list the instructions for all your procedures in your diary. Write down your protocols in detail *before* you carry them out, see what you change when you actually do the experiments, and then correct the original protocols. In addition, derive formulas and make computations directly in the *Materials and Methods* section of your records.

Get in the habit of writing notes during and after each experiment. Remind yourself of particular events and times. Note all the problems, changes, and questions that come up during the experiment. And, remember to quantify whenever possible—measure or estimate times, locations, numbers, amounts, and sizes.

1.1.2.   Results

Your diary is the best place to record your data, so that it will retain its connections to concurrent events. When you make a data entry, record *all* your results, whether full, partial, incomplete, or erroneous. This will be the raw stuff of your paper's *Results* section, so make the notes detailed. Consider the raw data in your diary to be permanent records. You will refer to these as you write this paper, future papers, and future grants, and you may find it necessary to show these records to people who question your conclusions.

Get in the habit of adding commentaries about your data, including times, dates, amounts, errors, and impressions. Write, for example, "2:12 pm may have spilled 2–3 drops of solution A at edge of slide," "4:30 pm bumped table shaking the apparatus," "5:00 am needle inserted smoothly, no problems with injection," "11:52 am Mrs. Jones (subject D2) sneezed, causing a blip on the meter," and even, "3–5 am incredibly boring wait, dozed on and off but don't think I missed anything."

Again, make a special effort to measure everything. Records of your experiments should be filled with numbers—"2 home visits" "6 cm wide" "18 degrees to the left" "20 drops" "16.5 min" "3 tablets" "8 stripes" "5 repetitions" "2 cm beyond the second bar" "4 pushes of the lever".

1.2.   References—Archive Your Sources

Your second type of computerized notebook should be devoted to references. Every time you get information from a colleague, read a relevant article, find an interesting comment on the Internet, attend a useful seminar, or find a pertinent paragraph in a book, make a note in your reference record. Computerized reference records make these notes easy to organize, to search, and to reorganize. Later, when you need to piece together a bibliography, the software can put the references in the appropriate bibliographic format and sort them in the order that matches the citations in your manuscript.

Some bibliographic software packages can capture and record citations from databases and websites. Bibliographic software will also allow you to add notes and to attach files and images to each citation. Then, on demand, the programs can generate a bibliography (i.e., a selected list of references) in the format required by a particular journal.

Commonly used bibliographic software programs include *EndNote*, *ProCite*, and *Reference Manager*. The Center for History and New Media at George Mason University offers a free and easy to use program called *Zotero*, which can be downloaded from *http://www.zotero.org*

2.   BEGIN A DRAFT EARLY

Your research notebooks should be ongoing projects that start on Day 1. In addition to the notebooks, early in your research you should begin to put together a rough draft of your paper, well before your experiments are completed. Therefore, when you need a break from bench work, take an afternoon, go to your desk, and lay out the outlines of your paper.

Consider paper-writing to be an integral part of your research. As you translate your plans and records into words and as you draw diagrams and construct tables, you will see your activities in a clearer light. You will also gain a firmer control over your concurrent experiments.

Later, when your experiments are largely finished, you will still be writing and revising your manuscript. Undoubtedly, you will find holes in your notes. This is natural, it is inescapable. When you discover that something is missing or that procedures could have been done more thoroughly or more accurately, take a deep breath, put down your pen, and go back to the lab. Look in more detail, test additional subjects, repeat experiments if it is possible, and tie up loose ends.

Wrestling with your manuscript engages you in exploratory data analysis. You will learn from your data, and you will revise your ideas. Good research is cyclic and iterative, not linear and final. The process of writing may set things in a different light, and when this happens, don't hesitate to rework your work.

Chapter 2

COMPOSING THE SECTIONS
OF A RESEARCH PAPER

To deliver content with the least distractions, scientific papers have a stereotyped form and style. The standard format of a research paper has six sections:

- *Title* and *Abstract*, which encapsulate the paper
- *Introduction*, which describes where the paper's research question fits into current science
- *Materials and Methods*, which translates the research question into a detailed recipe of operations
- *Results*, which is an orderly compilation of the data observed after following the research recipe
- *Discussion*, which consolidates the data and connects it to the data of other researchers
- *Conclusion*, which gives the one or two scientific points to which the entire paper leads

This format has been called the IMRAD (Introduction, Materials and Methods, Results, And Discussion) organization. I,M,R,D is the order that the sections have in the published paper, but this is not the best order in which to write your manuscript. It is more efficient to work on the draft of your paper from the middle out, from the known to the discovered, i.e.

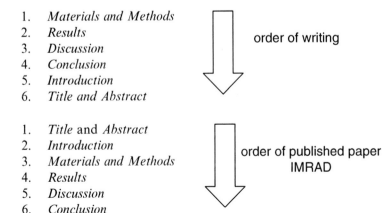

1. *Materials and Methods*
2. *Results*
3. *Discussion*
4. *Conclusion*
5. *Introduction*
6. *Title and Abstract*

order of writing

1. *Title* and *Abstract*
2. *Introduction*
3. *Materials and Methods*
4. *Results*
5. *Discussion*
6. *Conclusion*

order of published paper
IMRAD

The order of your writing should follow the order of your scientific analysis:

M.J. Katz, *From Research to Manuscript*,
© Springer Science + Business Media B.V. 2009

- Your *Materials and Methods* can be described before you have generated your *Results*.
- Your *Results* must be collected and organized before you can analyze them in your *Discussion*.
- Your *Discussion* recaps your *Results* and points you to a *Conclusion*.
- You must know your *Conclusion* before you can write an *Introduction* that sets the *Conclusion* in its natural place in science.
- The *Introduction* shows that your *Conclusion* was previously unknown or unproven.
- The *Title* and *Abstract*, which summarize your paper, must first have a completed paper to summarize.

1.   MATERIALS AND METHODS

> *Skeleton of the* Materials and Methods *Section*
> A. Recipe no. 1
> B. Recipe no. 2
> C. ...

Begin writing your paper with the *Materials and Methods* section. Your *Materials and Methods* give definition and meaning to your data. Scientific data are not absolute. They are contextual: they make sense only in the context of the procedures used to generate them. Therefore, you need to present your data in the company of detailed descriptions of your tools and complete instructions for your experimental procedures.

Results are only scientific when accompanied by the recipes used to generate them, and the *Materials and Methods* section is considered so fundamental a part of any research paper that it is the one section reviewers will rarely ask you to trim.

Recipes make results meaningful. In addition, recipes can be the essential science itself. In the last 30 years, almost half of the Nobel Prizes awarded in chemistry and in medicine were for the development of recipes—i.e., laboratory techniques and technologies.

1.1.   Recipes

1.1.1.   A *Recipe → Results* Report

At the end of the nineteenth century, scientific research papers began to set aside a separate section in which the authors described their experimental protocols. Today, all scientific papers are required to include a *Materials and Methods* section or its equivalent. As the *Uniform Requirements for Manuscripts Submitted to Biomedical Journals* states, the task of the *Materials and Methods* section is to

"Identify the methods, apparatus ..., and procedures in sufficient detail to allow other workers to reproduce the result."

This last mandate, "in sufficient detail to allow other workers to reproduce the result," is shorthand for an implied agreement between the researcher and the scientific community. This tacit covenant comprises three implicit statements by a paper's author:

- "In this paper, I specify the operations that compose a *recipe x*, which is described in a form that, I believe, can be performed by other researchers."
- "By following *recipe x*, I observed *results y*, which are detailed in this paper."
- "I predict that anyone else who carries out *recipe x* will also observe *results y*."

The last statement can be symbolized as *recipe x → results y*, and together, the set of three statements can be abbreviated as "this paper contains a *recipe → results* report."

The core of a scientific research paper is its *recipe → results* report. The recipe is described in the *Materials and Methods* section. The results are described in the *Results* section. Scientific papers contain other sections, such as an *Introduction* and a *Discussion*, but the irreducible core of a research paper is its central pair of sections, the *Materials and Methods* and the *Results*.

1.1.2. Repeatable Recipes and Reproducible Results

Nature is complex. We are surrounded by things that cannot be reproduced. For example, suppose that your hobby is flipping coins. One morning, you flip 12 pennies, and you find that they all land heads side up. This seems remarkable to you, and you decide to report it in a scientific paper.

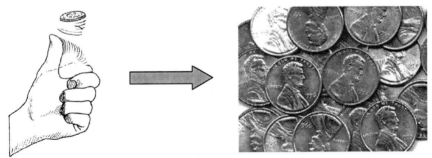

You sit down to block out your paper, and you begin by drafting the *Materials and Methods* section. You know that every scientific paper needs a trustworthy set of operational recipes that come with your tacit pledge, "To the best of my knowledge, if you follow my recipes, then you will get my results."

To make good on this promise, you must offer operational details with sufficient clarity that other researchers will be able to reproduce your experiments. Your *Materials and Methods* section must describe exactly how to grow the same tissue-culture axon patterns that you have grown, it must give blueprints for the forge and the wire-puller that will spin the same flexible, high-conduction beryllium wire that you have spun, or it must lay out the maps that will unerringly lead someone else to the yeti colony that you have found. The critical principle is:

- Every observation recorded in the *Results* section of your paper must be the product of repeatable procedures—practicable recipes—that are completely detailed in the *Materials and Methods* section.

When you were flipping 12 Heads in a row, you were not aware of doing anything that you had not done before, and to check, you again flip 12 coins by following your standard procedures and this time, you get 5 Heads and 7 Tails. To write the *Materials and Methods* section of a scientific paper about flipping 12 Heads in a row, you need to detail a recipe that others can repeat to reproduce your results. You realize that you do not have such a recipe, so you set aside the idea of writing a research report on your experience.

1.2.   Writing Your *Materials and Methods* Section

1.2.1.   Your Daily Lab Notebook Is a First Draft

As you compile your experimental records, you are writing the first draft of your *Materials and Methods* section. Be sure to record your materials and methods

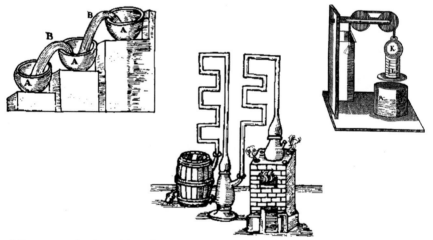

while you are in the midst of your experiments, when all the technical details are still fresh in your mind. List all the substances and supplies, the tools, instruments, appliances, contrivances, techniques, procedures, and solutions. Write full recipes, and draw diagrams.

Record every single item and operation you use, and include full details. For example:

- "Used fire-polished borosilicate glass pipette. Placed single *Xenopus* embryo, pre-washed and without vitelline membrane, in autoclaved glass petri dish containing 10% Holtfreter's solution made with double distilled tap water. Repeated 7 times."

As you take notes, record even seemingly inconsequential things, such as,

• "Wore sterile gloves, used small autoclaved glass spoon"

Later, you will weed out any unnecessary details.

1.2.2. Be Exhaustive

Remember—your goal is completeness.
For the *Materials*, your records should:

• Specify the apparatuses, including the manufacturers' names and addresses.
• Identify all chemicals and supplies used, and for drugs, include the generic names, doses, and routes of administration.
• Describe the salient characteristics of all experimental subjects and tell exactly how the subjects were found or chosen, including the eligibility requirements, the criteria for exclusion, and the nature of the source population from which they were drawn.
• Define everything in practical, operational terms—for example:
    o "age was ascertained by asking the person" or "age was ascertained by checking the person's driver's license"
    o "temperature was measured under the tongue using a flexible-tip standard digital hospital thermometer (LifeSource – A&D Medical)" or "temperature was measured in the ear using a BV Medical Instant Ear thermometer"

For the *Methods*, your records should:

• Explain the overall design of your research program
• Fully describe all the operations and procedures in sufficient detail to allow other workers to repeat them and to reproduce your results
• Cite references for all previously documented methods, including any statistical methods
• Give complete recipes for any new or modified techniques
• Explain the procedures used to analyze your data

Chocolate Pound Cake

3 cups sugar
1/2 cup vegetable shortening
1 cup milk
5 tsp. cocoa
1 tsp. vanilla extract

1 cup unsalted butter
5 eggs
3 cups plain flour
a pinch of salt

Cream together sugar, vegetable shortening, and butter. Add eggs one at a time. Sift together dry ingredients and add to mixture alternately with milk.

Coat a tube cake pan with vegetable shortening and then closely line with wax paper. Pour cake mixture into cake pan and bake at 325 degrees for approximately one hour and ten minutes. Test for doneness with a wooden toothpick (Granny always used a bristle from the broom but a toothpick is cleaner). The toothpick should come out dry. If not, bake for another ten minutes.

When done, allow cake to cool for about 20 minutes and then invert the cake pan on a cake plate and remove the wax paper from the cake.

1.2.3.   Use Algorithms

Most of the *Materials and Methods* section of your final paper will take the form of recipes, and the archetypal form for describing a recipe is the *algorithm*. An algorithm is a well-defined program describing the steps necessary to get from Point A to Point B.

From the beginning, when recording your recipes in your diary, write algorithms. For instance, don't write:

• "The location of the axon tip was measured accurately every 10 min."

This sentence does not fully describe a repeatable procedure: it is not a detailed algorithm. Instead, tell us exactly what you did; for example,

• "At 10 min intervals, the videotape was stopped. The location of the axon tip was then measured (within an error of +/−1 grid square) on a 15 cm × 15 cm transparent plastic grid of 5 squares/cm."

Likewise, do not write the instructionless

• "Fractal dimensions were tested for their statistical significance."

Instead, write down the complete set of instructions that will allow other researchers to repeat your actions:

• "*P* values for statistical significance were calculated from t-tests of the means of the logs of the fractal dimensions. (See the *Appendix* for our method of calculating fractal dimensions.)"

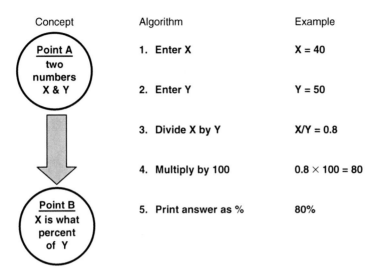

|| Concept | Algorithm | Example |

Concept	Algorithm	Example
Point A two numbers X & Y	1. Enter X	X = 40
	2. Enter Y	Y = 50
	3. Divide X by Y	X/Y = 0.8
	4. Multiply by 100	0.8 × 100 = 80
Point B X is what percent of Y	5. Print answer as %	80%

Algorithm: **Mechanical Instructions for Getting from Point A to Point B**

The *Materials and Methods* must be detailed and unambiguous. It is helpful to imagine that the algorithm you are writing is a computer program in which you can give only binary ("yes/no" "on/off" "always do *x*") definitions and instructions.

A computer cannot understand imprecise words, such as 'sometimes', 'on occasion', or 'maybe.' When you write:

- "I used a Teflon-coated glass spatula"

you must mean that you *always* used a Teflon-coated glass spatula. If that was not the case, don't write, "Sometimes, I used a Teflon-coated glass spatula" or "At times, it was necessary to use a Teflon-coated glass spatula." Be specific, for example,

- "When the temperature was below 37°C, I used a Teflon-coated glass spatula."

Even the scientific-sounding term 'approximately' does not belong in a *Materials and Methods* section, so don't write, "Measurements were taken at approximately 45 mph." Instead, write, "Measurements were taken when the digital speedometer read 45 mph" or "Measurements were taken when the dial on the speedometer read 45 mph +/− 1 mph."

Also, be sure to define all your technical terms and abbreviations when they first appear, e.g., "Cortef (Pfizer brand hydrocortisone tablets). ..."

The thoroughness required for writing algorithms makes for tedious prose, and your sentences will be formulaic. A typical *Materials and Methods* recipe might include:

- "Twelve ml of normal saline was flushed through the cannula. The cannula was allowed to drain for 1 min. Another 12 ml of normal saline was then flushed through the cannula. Immediately thereafter, 50 ml of the test solution was flushed through the cannula. The cannula was again allowed to drain for 1 min ... "

This reads like an instruction manual—and that is exactly your goal.

1.2.4. Include Detailed Instructions

To illustrate the detail needed in the *Materials and Methods* section, here are excerpts from a variety of scientific papers.

- From a report on the role of juvenile hormone during the development of moths (Williams, 1961):

"**2.1.   Excision of larval corpora allata**

"Matched pairs of larvae were sacrificed and the corpora allata dissected from their heads, as previously described in the case of adult corpora allata (Williams, 1959, page 327). In certain experiments, it was necessary to excise the corpora allata without killing the larval donors. By adaptations of a method suggested by Dr. William Van der Kloot (unpublished observations), a surgical approach through the underside of the neck was utilized, as follows:

"A 3-cm. length of dowel was attached to a small base-board so that the dowel stood vertically above the perforated plate of the anesthesia funnel. The top end of the dowel was grooved to fit the dorsum of the larval head capsule. The larva was deeply anesthetized and placed in the anesthesia funnel so that the underside of the neck was stretched and flexed over the top end of the dowel. The head capsule was held in the groove by small clips so that the thorax and anterior abdominal segments hung vertically. In this way, the blood was pressed from the neck region and the latter was flattened and essentially bloodless.

"The ventral midline of the neck was lifted with forceps and a single V-shaped incision was made through the integument with microscissors. Under a dissecting microscope the operation was carried out through the incision, first on one side of the neck and then on the other. With blunt probes, the muscles of the neck were pressed apart and the corpora allata located and excised.

"At the conclusion of the operation, the flap of skin was spread in place. The animal was stored in a capped cardboard container at 5°C. until the next day. It was then returned to room temperature and placed on a netted branch of wild-cherry leaves."

- From a report on the Kaposi's sarcoma-associated herpesvirus (KSHV) effects on infected human cells (Glaunsinger and Ganem, 2004):

"**mRNA Amplification, Microarray Hybridization, and Data Analysis**

"The human cDNA array has been described previously (8) and represents ~20,000 genes derived from PCR of an expressed sequence tag (EST) clone set using common primers as well as 200 additional cDNAs amplified using gene-specific primers for KSHV sequences and select additional cellular genes. To generate mRNA for microarray analysis, 5 μg of total RNA from each sample was subjected to linear mRNA amplification by in vitro transcription of cDNA as described previously (13). To generate each probe, 2 μg of the amplified RNA was reverse transcribed in the presence of 300 μM amino-allyl dUTP as described previously (14) and coupled to either Cy3 or Cy5 (Amersham Biosciences). Probes were hybridized to the microarrays overnight at 65°C. Arrays were scanned using an Axon 4000B scanner, and Cy3 and Cy5 signals were normalized such that all good features (r2 ≥ 0.75) equaled 1. Arrays were analyzed using GenePix 3.0 and clustered using TreeView. Spots with obvious defects were excluded from the analysis. Each experiment was repeated three independent times, and genes up-regulated at least twofold greater than the reference sample in two or more of the experiments were considered significant. The fold up-regulation reported in Tables I and II represents the average fold up-regulation of that gene from the set of experiments. The complete array datasets can be viewed on the NCBI-GEO web-site (www.ncbi. nlm.nih.gov/geo, accession no. GSE1406)."

- From a report on using high doses of cytosine arabinoside (ARA-C) to keep patients with acute myeloid leukemia in continuous complete remission (Bohm et al., 2005): (CALBG = Cancer and Leukemia Group B, HiDAC = high dose ARA-C)

"2.2. Treatment schedules
"Remission induction treatment consisted of daunorubicin (45 mg/m²/day i.v., days 1–3), etoposide (100 mg/m²/day i.v., days 1–5), and ARA-C (2×100 mg/m²/ day i.v., days 1–7 = DAV, 3 + 5 + 7) [25]. In case of blast cell persistence, patients received a second cycle of induction chemotherapy. In most patients the second cycle consisted of ARA-C, 2 × 1,000 mg/m²/day, days 1–4 and mitoxantrone, 12 mg/m²/ day, days 3–5 (MiDAC), whereas three patients received DAV as second induction cycle. In three patients, a third induction cycle consisting of fludarabine, 30 mg/m²/ day, days 1–5 and ARA-C, 2,000 mg/m²/day, days 1–5, and G-CSF, 300 μg s.c. per day until neutrophil recovery (FLAG) was administered. Patients with blast cell persistence after three induction cycles were excluded. Consolidation consisted of 2 × 3 g/m²/day ARA-C i.v. (3 h-infusions in 12 h-intervals) on days 1, 3, and 5 (HiDAC), with a total number of four cycles of HiDAC aimed to be administered according to the protocol described by the CALGB study group [16]."

- From a report measuring the interface tensions in immiscible fluid mixtures as the fluid boundaries disappear (Sundar and Widom, 1987):

"Density Determination
"The densities of the coexisting phases in each of the three-phase mixtures were measured by pycnometry. The pycnometer used was a variation of the Ostwald-Sprengel type and is shown in Figure 3. It consisted of a flat-bottomed glass bulb of about 20-cm³ volume, with two arms of sections of 1/4-in. o.d. glass tubing, each of which had incorporated in it a short section of heavy-walled precision bore capillary tubing of 0.01-in. diameter. The open ends of the pycnometer limbs were sealed when needed by using 1/4-in. Swagelok end caps with Teflon ferrules. The volumes of the pycnometers up to a pair of calibration marks on the capillary arms were determined by using boiled and filtered distilled water at the temperatures of interest.
"The pycnometer, once suspended in the water bath, was filled as follows. One end of a length of 1/16-in.-diameter Teflon tubing was connected to one of its limbs

via a Swagelok reducing union. This Teflon tube was threaded through an aluminum tube, much of the length of which was covered with heating tape to permit the samples to be kept heated during the measurements at the higher temperatures to prevent phase separation. The other end of the Teflon tube was immersed in the phase whose density was being measured. The pycnometer was filled by applying gentle suction with a syringe with a large-bore needle attached to its other limb. After slightly overfilling, the syringe and tubing were disconnected and the contents allowed to regain thermal equilibrium. Enough liquid was then removed with a syringe so as to have the liquid levels in the capillary sections. By then measuring the heights of these levels above or below the calibration marks, the volume was accurately determined. The mass was obtained by difference and the density thus determined."

• From a report on the use of proton magnetic resonance spectroscopy to assess the severity of damage in a shaken infant (Haseler et al., 1997):

"Patients

"In three infants (aged 5 months [A], 5 weeks [B], and 7 months [C]) the diagnosis of SBS was established clinically, and in particular, severe bilateral retinal hemorrhages were identified.

"Infant A (5-month-old female) was admitted with apnea after a seizure that was reported initially as the result of a fall from a table (about 3 feet). However, one parent allegedly observed the other parent severely shaking the infant. There was no prior history of abuse, and no fractures were noted on skeletal survey.

"Infant B (5-week-old female) was admitted within 4 hours of the injury when the mother brought the infant to the hospital after returning home to find her behaving "abnormally" in the care of the father. Police records were strongly suggestive of shaking without blows, but there was no other history of abuse. Skeletal survey was negative and no evidence of chronic abuse was detected. A clinical examination 2 weeks postnatal was normal.

"Infant C (7-month-old female) was admitted comatose after falling down three stairsteps. There was never a clear history of abuse and therefore the interval postinjury is more difficult to assign. Initially, SBS was not suspected, but on skeletal survey, a fractured occipital condyle was noted. Evidence for prior injury was limited to a possible chronic subdural hematoma on MRI.

"The clinical and neurological status of each infant at time of admission and on the days MRS was performed is shown in Table 1. Clinical status and outcome after long-term follow-up are also presented in Table 1."

• From a report on the aging changes of Au/n-AlGaN Schottky diodes when exposed to air (Readinger and Mohney, 2005):

"Preparation of the AlGaN surface for all processing began with a standard solvent clean (acetone, methanol, and deionized (DI) water rinse for >1 min) to remove the carbonaceous contamination layer, then blowing the samples dry with N_2. The remaining portion of the contamination overlayer on GaN is presumed to exist as a surface oxide or oxynitride (24, 25). The native oxide on an AlGaN alloy has been shown to be a mixture of Ga_2O_3 and Al_2O_3 (26), but the oxide is rich in Al_2O_3 even at small Al fractions within the semiconductor (27, 28). Prior to metallization, the AlGaN surface was treated at room temperature for 10 min in a solution of either HCl (37%) diluted 1:1 with DI water or buffered oxide etch (BOE). Following the surface treatment, the samples were rinsed in DI water and blown dry with N_2. After surface preparation, the diodes were fabricated by fixing the samples beneath a shadow mask with 250 μm diameter holes and then placing them directly into the vacuum chamber for metal deposition. Following the deposition of the Schottky contacts, a sputtered or e-beam evaporated large-area Al pad was deposited over one quarter of the sample area and used as the ohmic contact."

- From a report on the effects of three natural agonists on the contraction of smooth muscle in lung airways (Perez and Sanderson, 2005):

"Lung Slices

"To preserve the normal morphology and study the physiological response of intrapulmonary airways and blood vessels, we modified the preparation of lung slices (Bergner and Sanderson, 2002a). Male BALB/C inbred mice (Charles River Breeding Labs, Needham, MA), between 7 and 9 wk old, were killed by intra-peritoneal injection of 0.3 ml of pentobarbital sodium (Nembutal) as approved by the IACUC of the University of Massachusetts Medical School. The trachea was cannulated with an intravenous (IV) catheter tube with two input ports (20 G Intima; Becton Dickinson) and secured with suture thread (Dexon II, 4–0; Davis and Geck) to ensure a good seal. A syringe filled with 3 ml of air was attached to one port while the other port was closed. The chest cavity was opened by cutting along the sternum and the ribs adjacent to the diaphragm. To reduce the intrapul-monary blood vessel resistance and facilitate vessel perfusion with gelatin, the collapsed lungs were gently reinflated to approximate their total lung capacity by injecting 1.5 ml of air. A warm (37°C) solution of gelatin (type A, porcine skin, 300 bloom, 6% in sHBSS) was perfused through the intrapulmonary blood vessels, via the pulmonary artery, by inserting the hypodermic needle of an infu-sion set (SV × S25BL; Terumo Corporation) into the right ventricle of the heart and slowly injecting 1 ml of gelatin solution. A small cotton–wool swab soaked in ice-cold sHBSS was placed only on the heart to solidify the gelatin before the perfusion needle was removed.

"The lungs were deflated by releasing the positive air pressure. A syringe filled with a warm (37°C) solution of 2% agarose (type VII or VII-A: low gelling temperature) in sHBSS was attached to the second port of the catheter. The IV tube was clamped proximal to the trachea and purged of air with the agarose solution by allowing the trapped air to escape via a 27-gauge needle inserted into the IV tube proximal to the clamp. The IV clamp was removed and the lungs were reinflated by injecting 1.3 ml of agarose-sHBSS. Subsequently, 0.2 ml of air was injected into the airways to flush the agarose-sHBSS out of the airways and into the distal alveolar space. Immediately after agarose inflation, the lungs were washed with ice-cold sHBSS, and the animal was placed at 4°C for 15 min. The lung and heart were removed and placed in sHBSS (4°C) and cooled for an additional 30 min to ensure the complete gelling of the gelatin and agarose."

1.2.5.   Write a *Statistical Methods/Experimental Plan* Subsection

If your experiments produce numerical data, you will undoubtedly describe the data statistically. The *Materials and Methods* should explain the statistical methods you have used. In your *Statistical Methods* subsection, give the definitions of statistical terms, abbreviations, and symbols that you use and include citations of your exact sources, i.e., statistical books, technical papers, or computer software. Your goal is to give sufficient detail for a knowledgeable reader to repeat your statistical analyses and to reproduce your results.

When you have built your experiments using a preplanned statistical design, the *Materials and Methods* section is the place to explain the details. Randomized control trials are a common experimental design, and the *CONSORT Statement* is an excellent guide for organizing reports of this variety of experiment. (The *CONSORT Statement* can be read at and downloaded from *http://www.consort-statement.org/?o=1001.*)

As examples, here are the *Statistical Methods* subsections from three papers.

• From a report examining the ultrasound echogenicity of the substantia nigra in Parkinson's disease (Berg et al., 2001): (PD = Parkinson's disease, SN = substantia nigra):

"Statistics
"Descriptive statistics are given as median with lower and upper quartiles (25th and 75th percentile respectively). Results of SN echogenicity of PD patients were compared with measurements of hyperechogenic areas at the SN of 30 age-matched controls examined by the same sonographer with the same ultrasound system [3]. The upper standard deviation of SN echogenicity in the controls group was used as the cut-off for further analyses. Intergroup comparison was performed by the Mann-Whitney U-Test. Correlation analysis was performed by Spearman rank correlation."

• From a report on using the tri-block polymer P188 6h after a spinal cord injury in a mammal to increase the function and heal the structure of the spinal cord (Borgens et al., 2004):

"Statistical Evaluation
"Statistical computations were carried out using InStat software (GraphPad, San Diego, CA). Comparison of the proportions of animals in each group tested for evoked potentials was carried out using Fisher's exact test (two-tailed) and a comparison of means with Mann-Whitney nonparametric two-tailed test. Normalized measurements from 3D reconstructions of control and experimental groups were compared using unpaired, two-tailed, Student's t-test."

• From a report examining the effect of matrix metalloproteinase inhibitors on healing after periodontal surgery (Gapski et al., 2004): (AFS = access flap surgery, BOP = bleeding on probing, CAL = clinical attachment levels, LDD = low dose doxycycline, PD = probing depth):

"Statistical Analyses
"Data available for each patient were subjected to an intent-to-treat analysis and included full-mouth clinical measurements for three different parameters

(PD, CAL, and BOP) examined at six sites per tooth. Sample size determination for this pilot study was determined from Golub et al., (8) using 80% power for differences expected in a primary biological outcome measure, the bone resorption marker, ICTP. The level and prevalence of 40 species and ICTP levels (pg/site) was measured from the surgically treated quadrant excluding third molars. Clinical, microbial, and ICTP data were averaged within a patient and then compared between groups. In other analyses, data were stratified according to baseline PD of 1 to 4, 5 to 6, and ≥ 7 mm and averaged in a patient and then across the study population for clinical parameters and ICTP levels. Differences between drug and placebo groups were performed using the Mann-Whitney test. Differences with a P value less than 0.05 were considered significant. Microbiologic data were analyzed separately for sites receiving AFS+LDD and AFS+placebo. Mean levels (counts $\times 10^5$) for each of 40 species were computed for each patient in each treatment category at each visit. Significant differences over time were determined separately for each treatment category using the Quade test and the differences between groups were determined by the Mann-Whitney test. All analyses were performed using the patient as the unit of analysis."

1.2.6. Organize the *Materials and Methods* as an Instruction Manual

The *Materials and Methods* is the instruction manual for everything you have done in your research project. Give your instruction manual an overall organization that is easy to understand.

First, group the instructions together into units that contain complete recipes. Label these groups with clear, explanatory titles, such as:

- *Anesthetizing the Fruit Flies*
- *Building the Electrodes*
- *Calculating the Surface Areas*
- *Calibrating the Oscilloscope*
- *Fitting Caliper Bars to Subjects*
- *Interpreting the Chromatographic Scans*
- *Measuring the pH of the Initial Samples*
- *Normalizing the Tonal Response*
- *Preparing the Staining Solutions*
- *Recruiting the Volunteers*
- *The Steps in the Transplant Surgery*
- *Statistical Methods*

Next, organize the groupings so that readers can quickly find whichever set of instructions they are looking for. Often, the *Materials and Methods* recipes are presented chronologically, in the order that they are done during the experiment. You may, however, feel that it is clearer to group them some other way. For instance, you may decide to put the chemical recipes together in one subsection and the procedures dealing with experimental subjects in a separate subsection. There is no one best organization, but you should pick a scheme that makes your instruction manual easy to navigate.

1.3.   An Example of a Complete *Materials and Methods* Section

"Materials and Methods

Cells

Neurons were stained in histologic sections of tadpoles and in tissue culture preparations of frog and chick embryonic cells.

Xenopus tadpoles were raised from laboratory colonies. Primary amphibian tissue cultures 24–48 h old were grown by disaggregating neural tube cells of tailbud stage Xenopus embryos and plating the cells on glass coverslips in a modified Niu-Twitty solution (Hamburger, 1960) with 2% fetal bovine serum, nerve growth factor (NGF), and antibiotics added (adapted from Spitzer and Lamborghini, 1976).

Primary avian tissue cultures 24–48 h old were grown by disaggregating dorsal root ganglion cells of 7–12 d chick embryos and plating the cells on glass coverslips in Leibovitz's (L-15) medium (GIBCO) with 10% fetal bovine serum, 0.6% glucose, 0.3% methyl cellulose, NGF, antibiotics, and cytosine arabinoside. (Further details can be found in Shaw and Bray, 1977).

Histologic preparation

a) Tadpole sections

Whole tadpoles (Xenopus stage 48—Nieuwkoop and Faber, 1967) were fixed by immersion for 45 min in an ethanol fixative (90 ml 80% ethanol, 5 ml formalin, 5 ml glacial acetic acid, 0.035 g NaCl, 1 ml DMSO). The tissues were then washed, embedded in paraffin, and sectioned at 12 µm. Slides were deparaffinized and hydrated to glass-distilled water.

b) Cultures

Cell cultures were fixed by flooding them with the same ethanol fixative. The first fixative bath was immediately replaced with fresh fixative, and, in this, the cells (on glass coverslips) were allowed to continue fixing for 45 min. Finally, the coverslips were rinsed ten times in tap water.

Staining

a) Bodian stain

Slide and coverslip preparations were stained according to the following schedule:
1. Immerse in solution **A** for 24 h at 37°C.
2. Rinse well and soak in solution **B** for 10 min.
3. Rinse well and soak in solution **C** for 1 min.
4. Rinse well and soak in solution **D** for 3 min.

(continued)

continued

---

5. Rinse well and soak in solution **E** for 7 min. Rinses were in three changes of glass-distilled water. Steps 2–5 were done at room temperature.

Bodian Solutions: (each made in 100 ml glass-distilled water)
    A = 1 g silver protein (Roboz Surgical Inst Co), 5 g copper foil
    B = 0.5 g potassium metaborate, 1 g hydroquinone, 5 g anhydrous sodium sulfite
    C = 1 drop glacial acetic acid, 1 g gold chloride
    D = 1 g oxalic acid
    E = 5 g sodium thiosulfate

b) Post-staining intensification

After the initial staining, slides and coverslips were soaked in intensifier solution **J** for 1 min, again in solution **C** for 1 min (for thick sections) or 10 min (for fixed cultures), then dehydrated and mounted with Permount.

Recipe for intensifier solution **J**: Add solution **F** to solution **G**. Stir thoroughly until no more precipitate forms. Add solution **H**. Stir until no more precipitate dissolves. Add solution **I** and stir well. Filter through Millex-GS 0.22 μm filter unit (Millipore Corp) to remove remaining precipitate.

Intensifier Solutions:
    F = 1.5 g silver nitrate, 25 ml glass-distilled water
    G = 1.5 g sodium sulfite, 25 ml glass-distilled water
    H = 2.6 g sodium thiosulfate, 25 ml glass-distilled water
    I = 0.4 g sodium sulfite, 0.6 g Elon (Eastman Kodak), 75 ml glass-distilled water

[This intensification method is a variant of a standard photographic enhancement procedure. See Hodgman et al., 1960.]"

1.4. Commit to a Few Key Variables

Your *Materials and Methods* section should be comprehensive. It is an instruction manual for all the things you have done to produce operationally-defined data. The other parts of your paper, however, will focus on only two or three aspects of that data. A scientific paper can explore in depth only a few of the many bits of information and intriguing byways that you have seen during your project. Make this clear to yourself early in your writing. After drafting a comprehensive *Materials and Methods*, explicitly limit your manuscript by defining the focal experimental variables of your scientific report.

Before going any farther, list those few experimental variables that you will be studying. Label them your "key variables." To keep you on track, write the key variables at the head of the drafts of the *Results*, *Discussion*, *Title*, and *Abstract* sections of your paper.

For example, suppose that you are evaluating the heights of 2-year-old girls whose mothers have IQs of > 130.

A thorough *Materials and Methods* for this study might include recipes for:

- *Finding High-IQ mothers of 2-year Old Girls*
- *Enlisting Mothers and Daughters in the Study*
- *Verifying IQ Scores*
- *Verifying Children's Ages*
- *Measuring Children's Heights*
- *Statistical Methods*

The *Materials and Methods* in the final paper should give complete instructions for all of these operations. From this comprehensive instruction manual, it may not be immediately apparent that the focus of your paper will be just two specific items. After composing a draft of your *Materials and Methods*, it will help you to focus your paper if you take a moment to list the two key experimental variables—the IQ of a mother and the height of her 2-year old daughter—before you dive into the *Results* section of your draft.

A key variable is an experimental variable, and it denotes the *type* of data on which you will focus when analyzing your experiments. Your *Results* section should present *all* the data values of that type produced during your research.

2.   APPENDIX

> *Skeleton of an* Appendix
>
> A. Title
> B. Long Recipe
> C. References

An *Appendix* is a self-contained addition to the *Materials and Methods*, although the *Appendix* is put separately at the end of the paper. In a scientific paper, an *Appendix* is not a commentary or an adjunct to the *Results* or the *Discussion*—it is a detailed explanation that is too long for the *Materials and Methods* section.

An *Appendix* might contain, for example, a long recipe for a chemical preparation. It might explain a mathematical formula, detail a computer program, or diagram the wiring or construction of an apparatus, such as this:

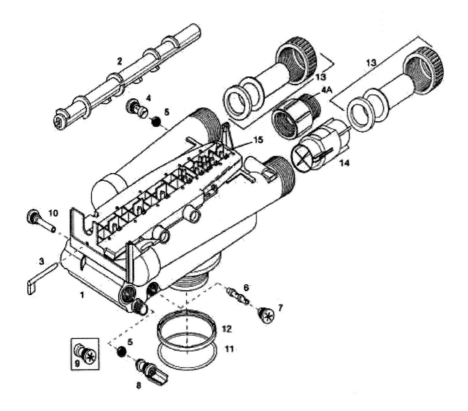

An *Appendix* could also illustrate a surgical operation used in your experiments, or it could reproduce the complete survey form used in collecting data.

Appendices are lettered, and they appear after the *References* section in a paper. The *Appendix* has a title and is a stand-alone entity. This means that if an *Appendix* includes bibliographic citations, then those citations are listed at the end of the *Appendix*, not in the *References* section of the main paper.

Here is an example of a complete *Appendix*.

"APPENDIX A

Calculating the Fractal Dimension of the Growth Path of an Axon

Topologically, a curve in a plane always has a dimension of 1. Nevertheless, as it becomes more and more convoluted, a curve fills more and more of the plane. From this perspective, a convoluted curve might be considered to have a fractional geometric or non-topological dimension – a "fractal dimension" – of greater than 1 (Mandelbrot, 1977, 1983). In terms of fractals, a straight line has a dimension of 1, an irregular line has a dimension of between 1 and 2, and a line that is so convoluted as to completely fill a plane has a dimension approaching the dimension of the plane, namely a dimension of 2. In this way, fractal dimensions assign numbers to the degree of convolution of planar curves.

The general form of a fractal dimension D of a planar curve is:

$$(\text{length})^{1/D} = K(\text{area})^{1/2}$$

where 'length' signifies the total geometric length of the curve, 'area' is the maximum potential geometric area that the curve could fill, and 'K' is a constant (Mandelbrot, 1977, 1983). For planar curves that are constructed of connected line segments, a practical approximation to this equation is:

$$D = \log(n) / \log(nd/L) = \log(n) / (\log(n) + \log(d/L))$$

where 'd' is the planar diameter of the curve (here estimated as the maximum shortest distance between any two line segment endpoints along the curve), and 'L' is the total geometric length of the curve (the sum of the lengths of all the line segments). This formula has the following limiting values:

- When the curve is a straight line, L is equal to the planar diameter of the curve, and the fractal dimension is D = 1.
- When the curve is a long random walk, L will be approximately equal to $(n^{1/2})$ L, and the fractal dimension will have D approaching 2.

Further details can be found in Katz and George, 1985.

Katz MJ, George EB. 1985. Fractals and the analysis of growth paths. *Bull Math Biol* 47: 273–286.

Mandelbrot BB. 1977. *Fractals: Form, Chance, and Dimension*. WH Freeman, NY. Mandelbrot BB. 1983. *The Fractal Geometry of Nature*. WH Freeman, NY."

3.   RESULTS

Skeleton of the Results *Section*

A. General Observations
B. Specific Observations
C. Case Studies
    1. Best Cases and/or
    2. Representative Cases

When on the run, scientists read the *Title* and the *Abstract* for a quick taste of a research paper. With more time, they also skim the *Introduction*, glance at the figures, and read the *Conclusion*. To study the article, they will look at the figures in the *Results* and read the *Introduction* and the *Discussion*. We all read the scientific literature this way, short-changing the innards of a paper and attending to the glitter at the beginning and the end.

In reality, however, while the edges may have the shine, the middle has the enduring substance. The glitter that is in a *Discussion* section is especially ephemeral, because it is in the *Discussion* that data is tied to scientific theories, hypotheses, and conjectures. Scientific generalities, such as these, are under constant renovation. Simple generalities come, they are tuned and tweaked, and then they are gone, sometimes suddenly, in a revolution, as Thomas Kuhn pointed out in *The Structure of Scientific Revolutions*.

Tweaking of a Theory

Central Dogma of molecular biology (1958)
-- unidirectional flow of information --

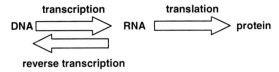

The change in the *Central Dogma* of molecular biology is a classic example of the life of a typical scientific theory. Originally, the biological rule was simple and stated that genetic information flowed only one way, from DNA to its products (RNA and proteins). This rule had a lifespan of about a dozen years, until it was shown that RNA from some viruses could restructure the DNA of an infected cell via reverse transcription.

The flashes and gleams of a scientific paper can be transitory, while the data reported in the paper usually last much longer. Carefully made scientific obser-

vations endure, and we can still use data from research papers more than a
century old.

The strongest part of a scientific research paper is at its center, in the *recipe*
→ *results* report. It is here that scientists tell their readers, "This is what you, too,
will find if you follow my recipes"—their recipes are given in the *Materials and
Methods* section, and their findings are described in the *Results* section.

3.1. Carving Out Your *Results* from Your Observations

3.1.1. Record Everything While You Experiment

While you were collecting your data, you should have been an immediate and complete
chronicler, putting everything you observed into your computerized research diary.

In the laboratory stage of your research, you should be recording things as
they happen, without trying to impose order or logic. Look at the focal point of the
experiment, but look also at the edges: write down the things that you planned to
study, and then note everything else that happens. And, whenever possible, write
down numbers—time things, count things, and measure things:

11 g   34 s   3,000 cpm   16.3 mg   106 m   1.5 flashes   0.441 psi

Your lab records should be a cluttered and unedited chronologic collection of
notes about what was happening during your experiments.

3.1.2. Limit Your Final Report to the Key Variables

The need to focus on a limited subset of your experimental observations is one of
the critical contributions that writing a paper makes to your research. As you begin
to draft your *Results* section, you should first select from your lab records only the
data concerning your key experimental variables. (The key variables, you will recall,
are those particular experimental observations that you chose to be the focus of your
paper when you finished blocking out a draft of the *Materials and Methods* section.)

From your clutter of data, you will have to set aside some of your experimental
observations, perhaps even some observations that fascinate you. For example, a
scientifically sound paper will report only complete data sets, so you should not
include tidbits that have not yet been fully explored and well documented, no matter how intriguing they may be.

In my paper on axon staining, I had to resist writing, "In preliminary
experiments, we found that the same stain intensifier worked well for fixed
sections of two fetal human nervous system samples." Even the tantalizing
"human result" is not appropriate in a research paper that is focused elsewhere
unless the data are:

(continued)

(continued)

> • Preceded by complete recipes detailing how the samples were obtained, treated, and analyzed
> • Complete and thoroughly described
> • Specific examples of the key variables of this particular paper

3.1.3.   Report All Data Produced by Following Your Recipe

You must limit your results to the key variables. If you are allowed to be selective, however, how much choosing can you do? In other words, how selective can you be when reporting your data?

For some perspective on this question, let me return to the coin flipping experience that we considered earlier, in the section on *Materials and Methods*. In that scenario, you had just flipped 12 coins and all the coins had landed heads side up. However, when you tried to write a scientific paper about your experience, you were stymied because you could not find a recipe that other researchers could follow to reproduce your results.

After thinking more about it, you realized that there *was* a recipe that you could offer. It was a recipe that other people could repeat and that would produce 12 consecutive Heads. This recipe was simply the exact set of steps that you had followed, because you knew that anyone who followed your recipe should eventually find that all 12 coins would one day turn up Heads.

Therefore, you again start writing a research paper. For your *Materials and Methods* section, you begin transcribing your recipe. First, you record the actual steps that you had taken:

• I began my experiment on April 16, 1983.
• I took 12 pennies from the change accumulated over a month's regular shopping.
• I sat down at my kitchen table at 8:00 am and flipped all 12 coins, one at a time.
• After all the coins had been flipped once, I counted the total number of Heads and the total number of Tails and recorded the numbers in my notebook alongside an entry of the date.
• I repeated the last two steps every morning, until one day—September 7, 2009—the procedure generated 12 Heads and 0 Tails. At this point, I stopped the experiment.

Next, you extract the essential algorithm:

1. Take 12 used pennies.
2. Flip them all, one at a time.
3. Repeat every day.
4. Stop when 12 Heads have been flipped during a single trial.

Because each coin flip is independent of the others, the chances of flipping 12 Heads in a row are 1 in $2^{12} = 4,096$, and every morning the chances are the same, 1/4,096. From probability theory, you calculate that there is a 99% chance that anyone following your recipe will get 12 Heads one morning if they are willing to try flipping coins for 18,860 days (51 years). (You were fortunate, because you chanced onto a 12 Heads morning in only 9,642 days.)

Although 51 years of daily coin flipping is a long time, you figure that a 99% chance of getting your result—12 heads during a single flipping session—qualifies your recipe as repeatable by any patient researcher. Therefore, you use your recipe as the basis of the *Materials and Methods* section of a scientific paper.

Now, you begin to draft the *Results*. What are your results? The results you would like to report are 'One morning, I flipped 12 Heads in a row.' Let's see whether these are the results of following your recipe.

First, we remind ourselves of the basic definition of a scientific observation.

• For a set of results (i.e., an observation) to be scientifically sound, it must be the complete and unadulterated output produced by strictly following an explicit recipe.

Now, we ask, is 'One morning, I flipped 12 Heads in a row' a scientific observation, i.e., is it the complete and unadulterated output of the recipe:

1. Take 12 used pennies.
2. Flip them all, one at a time.
3. Repeat every day.
4. Stop when 12 Heads have been flipped during a single trial.

To answer the question, you look at your lab notebook, which contains the complete output of your recipe. Your records—the *complete* results of following your recipe—are these 9,642 entries:

April 16 1983	HTHTTHTTTHTH
April 17 1983	HHTHTTHHHTTH
April 18 1983	TTHTHTTHTTTHT
April 19 1983	HTHTHTTHTHHT
...	
...	
...	
September 6 2009	THHTTTHHHTHTT
September 7 2009	HHHHHHHHHHHH

It is clear that 'One morning, I flipped 12 Heads in a row' is a minor component of your complete results. This one entry caught your attention, but, to be honest, you carved it carefully out of your long list of results, selecting this one specific sequence because it matched a particular pattern that interested you, namely, all Heads.

Stepping back, you realize that even if someone repeats your recipe, they will not reproduce your results. The chances are very small that they will reproduce the same sequence of 9,642 entries that you recorded in your research notebook.

The answer to the question, "How much selecting can I do when reporting my data?" is "Not much." For a research paper, you can only select the *type* of result to report. You are allowed to select which experimental variables are going to be central in your report. However, once you decide on a key experimental variable, you are obligated to report the complete data set generated by the recipe for that variable.

When you choose a key variable, you are actually choosing the *recipe* that produces that experimental variable. For a scientific research paper, you first select a recipe, and you then report its complete, unedited output.

3.2.   Exploratory Data Analysis

At this point, therefore, you should gather the complete output of your research recipe(s). Before you write any text for your *Results* section, you need to organize this data, and in this case, 'organize' means 'find a natural arrangement that will help your reader to see some of the structure inherent in the set of data.'

3.2.1.   Try Out a Variety of Visual Arrangements

People understand orderly things best. We need patterns and symmetry, regularity and *taxis*, harmony and *lucidus ordo*, i.e., bright clean order. Therefore, help your readers by finding the *lucidus ordo* in your data. Start without the preconceived idea that your data will have a certain particular order. Instead, work empirically, arranging and rearranging your data and looking for any inherent patterns.

In the *Results* section, your goal is to present your data with organization but without interpretation. Of course, each way that you organize your data will be based on some amount of interpretation; nonetheless, your goal is to have the data drive the arrangement. Therefore, as you search for a good organization, avoid squeezing your data into a pattern that you think should be there. Do your best to forget how the data ought to look or how you would like the data to look. Undoubtedly, you began your research with an idea of what to expect. Set this idea aside. Instead, let your data lead you.

Begin your search by taking your research records and spreading the full range of data on your desk.

As a sample data set, imagine that you have been studying the number of stripes found on Guatemalan tarantulas. In your field records, you labeled the tarantulas by day of collection (using letters) and order of discovery (using numbers). From your notebook, your records of tarantula (male and female *Brachypelma mythicosm*) observations were:

Day 1
A1: 7 cm female, 8 stripes
A2: 15 cm female, 9 stripes
A3: 14 cm male, 9 stripes
A4: 7 cm male, 7 stripes

Day 2
B1: 5 cm female, 8 stripes
B2: 9 cm male, 8 stripes
B3: 17 cm female, 7 stripes
B4: 13 cm female, 9 stripes

Day 3
C1: 11 cm female, 8 stripes
C2: 10 cm male, 6 stripes
C3: 12 cm male, 8 stripes
C4: 12 cm female, 6 stripes

Now do some exploratory data analysis. Shuffle the data into all possible combinations, looking for inherent patterns.

- Look at each key variable alone.
- Set variable against variable.
- Put the data in chronological order.
- Order the data by size, weight, gender, shape, color, or height.
- Try a variety of tables.
- If there is sufficient data, try histograms and other graphs.

3.2.1.1.   Tables
When you are exploring, make as many tables as you can.

With your tarantula data, for instance, you might try these differently-ordered tables:

CHRONOLOGIC ORDER			LENGTH ORDER			STRIPE ORDER			SEX ORDER		
length	sex	stripes	length	sex	stripes	stripes	sex	length	sex	stripes	length
7	F	8	5	F	8	6	M	10	F	6	12
15	F	9	7	M	7	6	F	12	F	7	17
14	M	9	7	F	8	7	M	7	F	8	5
7	M	7	9	M	8	7	F	17	F	8	7
5	F	8	10	M	6	8	F	5	F	8	11
13	F	9	11	F	8	8	F	7	F	9	13
17	F	7	12	M	8	8	F	11	F	9	15
9	M	8	12	F	6	8	M	12	M	6	10
12	M	8	13	F	9	8	M	9	M	7	7
10	M	6	14	M	9	9	M	14	M	8	9
11	F	8	15	F	8	9	F	13	M	8	12
12	F	6	17	F	7	9	F	15	M	9	14

3.2.1.2.   Graphs
If your tables are small, for example four or five entries, then the table by itself will probably be a fine presentation of your data. On the other hand, if your tables are large, then you can help your reader to see both the scope of your results and the inherent patterns by graphing the data. With big data sets, experiment with graphs.

For your spider data, you might draw this histogram, for example:

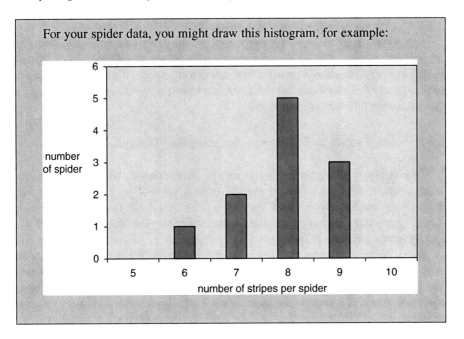

and this scatter plot:

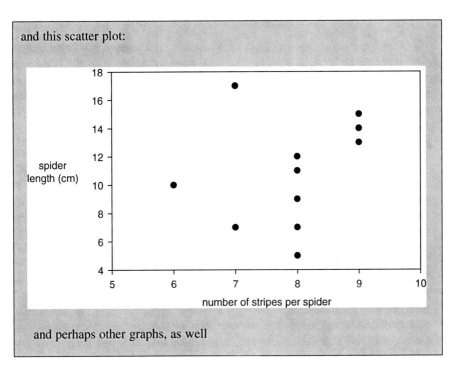

and perhaps other graphs, as well

3.2.1.3. Pictures, Diagrams, and Charts

Graphs are a good way to show visual patterns in numerical data. Because humans are so adept at understanding visual patterns, you should try to find a visual form for your data even when the data are not numbers. If your data are a set of photographs, for example, lay them out in a variety of arrangements, looking for patterns in the different presentations.

3.2.2. Pick a Simple Inclusive Arrangement for Presenting Your Data

Now compare all the arrangements that you have created. Your goal is find one arrangement that offers a comprehensive, orderly view of your data. During this exploratory analysis, you should be letting the data lead you—so, clear your mind, stand back, and pick one simple and satisfying presentation from all the arrangements you have devised.

It can be hard to argue that any one arrangement is the *best* way to organize your data. On the other hand, there are usually better and worse arrangements. The better arrangements are often emotionally satisfying. They tend to show neat, concise patterns that include the most data with the least squeezing, stretching, or trimming.

3.2.3. Write a Description of the Pattern in Your Data

You have chosen an organization for your data, and ideally, this organization can be represented by a picture, i.e., a visual representation. With the picture in hand, your next task is to describe it in words. Begin by studying the picture with a naive eye. Put aside any thoughts of the *meaning* of the data, just absorb the pattern(s). Then write a detailed description.

For instance, your description might be:

- "When the 500 (x,y) data points are graphed in a scatter plot (Graph 1), they form a peak with a sharp slope at low values of x and rising to a maximum at approximately $x = 50$. The curve of the scatter plot ends in a gently decreasing slope that tapers to a flat line beyond (approximately) $x = 250$. The greatest density of points is at the peak, between $x = 45$ and $x = 60$. The highest y value, which is located at the peak, is $y = 24.5$. No y value was less than $y = 0.6$ or greater than y = 24.5."

Or, for a different pattern, you might write:

- "Autoradiographs of the 21 gels could be divided into 3 distinct groups, as listed in Table 1. Group A contains 13 gels that had bands that co-migrated with actin but no bands that co-migrated with tubulin. Group B contains 5 gels that had some bands that co-migrated with actin and other bands that co-migrated with tubulin. Group C contains 3 gels that had bands that co-migrated with tubulin but no bands that co-migrated with actin. The 13 gels in Group A were from 2–3 day-old cultures, the 3 gels in Group C were from 4–6 day-old cultures, and the 5 gels in Group B were from 3–5 day-old cultures."

Now, set aside your picture and your description, and go back to your computer screen, where you can begin to write the main text of your *Results* section, which will put your data in its experimental context.

3.3. Writing the *Results*

Your *Results* section should have three parts, and each part will often be long enough to have its own heading.

- *General Observations.* The *Results* begins with a panoramic view of the research setting.
- *Specific Observations.* The *Results* then zooms in to focus on the data about your key variables, and it presents this data in the arrangement that you created during your exploratory analysis.
- *Case Studies.* The *Results* ends with one or two examples, showing the specific details of individual observations.

3.3.1. Subsection One—*General Observations*

When you write the *Results*, don't rush to the heart of your data. Ease your reader into the details by first giving a brief overview of your experiments in a *General Observations* subsection.

The *General Observations* subsection orients your reader. Here are some of the things you might describe:

- Features characterizing the population of experimental subjects.
  - o This type of introductory report might begin: "A total of 146 people filled out our questionnaire. Three individuals' forms were excluded because sections were illegible. The data were tabulated from the remaining 143 forms. As described in the *Materials and Methods*, all 143 subjects were male college students who reported their ages to be 19–23 yr old. Eighty-seven described their race as …"

- What would have happened if you had made no experimental manipulations or intrusions.
  - o This type of introductory report might begin: "Without intervention, the course of the infection was always …", "In normal mice, …", or "In the undisturbed natural setting, the oak leaves were …"

- The experimental setting.
  - o This type of introductory report tells the reader how things looked just before you began collecting data. This subsection might begin: "At the outset of the experiment, each subject was seated on a metal folding chair facing a blackened computer screen. …"

- A zoom-in view, beginning with a wide-angle shot.
  - o This type of introductory report might begin, "On opening the embryonic sac, the first things visible were …", "The answer sheets fell into two broad categories

…", or "All the control gels were stained either blue or red, and all the experimental gels had only yellow, yellow-green, or green bands. …"

Here are examples of *General Observation* subsections excerpted from the *Results* of a variety of papers:

- From a report on the role of juvenile hormone during the development of moths (Williams, 1961):

"When a living, active corpus allatum is implanted into a test pupa, the gland survives and becomes the site of synthesis and secretion of juvenile hormone. If several active glands are implanted, a corresponding number of synthetic centers are established. …

"Four (two pairs) adult corpora allata were implanted into abdomens of 24 test pupae; a second group of 24 pupae received only a single implant (one-half pair). The hosts were then placed at 25°C to await the onset of development.

"Three to four weeks later, a spectacular difference was evident between the two groups of animals. The individuals that received the four implants showed a generalized inhibition of adult differentiation, as signaled by the formation of a new pupal cuticle throughout broad areas of head, thorax, and abdomen. Indeed some of these animals could properly be described as second pupal stages in which only traces of adult characteristics had been differentiated. By contrast, the pupae that had received only a single corpus allatum ordinarily developed into adult moths and showed few abnormalities except for the formation of a new pupal cuticle along the plastic window where the pupal integument had been excised in the implantation procedure."

- From a report of growth of individual axons and accompanying sheath cells followed in the living animal over periods of days to weeks (Speidel, 1932):

"I. The normal distribution of nerves in the tail fin and the early changes

"The central portion of the tail of the tadpole is occupied by the spinal cord, axial skeleton (notochord), and the muscle mass. This region is not suitable for observation in the living animal. Emerging from between the muscle masses may be seen the nerves that supply the fin expansion. In general, these are segmentally arranged. Near the ventral and dorsal edges of the muscle mass longitudinal nerves may sometimes be seen which serve to connect adjacent spinal nerves with each other. Peripheral anastomoses between adjacent nerves also occur frequently. There is a set of nerves for the left side and a corresponding set for the right side. In the dorsal fin of older tadpoles halfway between the muscle mass and the fin edge extends a rather large longitudinal nerve, the dorsal branch of the ramus lateralis vagi."

- From a report consolidating evidence of the speed of extinction of North American dinosaurs (Fastovsky and Sheehan, 2005):

"In the latest Cretaceous of the North American Western Interior, dinosaurs such as Triceratops, Tyrannosaurus, and Edmontosaurus (and a host of lesser luminaries) roamed upland and coastal plain settings (Lehman, 1987) that formed during the Laramide phase of the Rocky Mountain uplift (Peterson, 1986). Dinosaur-bearing units that have been the subjects of studies sufficiently detailed to resolve the nature of the extinction are preserved in the structurally complicated Hanna Basin, an intermontane basin in southern Wyoming (Eberle and Lillegraven, 1998; Lillegraven et al., 2004), and undeformed sediments of the Williston Basin, an intracratonic basin extending through eastern Montana and western North and South Dakota (Peterson, 1986) (Fig. 1)."

- From a report on using high doses of cytosine arabinoside (ARA-C) to keep patients with acute myeloid leukemia (AML) in continuous complete remission (CCR) (Bohm et al., 2005):

"3.1.   Outcome of induction chemotherapy

"Seventy-three patients with de novo AML received induction treatment with DAV. Complete remission (CR) was obtained in 54/73 patients (74%). The majority of these patients (36/54 = 66.7%) entered CR after the first induction cycle, whereas in 18/54 patients (33.3%), more than one induction chemotherapy cycle (two cycles, n = 15; three cycles, n = 3) was required to achieve a CR. In 19 patients (26%), no CR could be obtained. In this group, 16 patients had blast cell persistence, and three patients died from treatment-related complications within 28 days after start of therapy (induction-related death)."

- From a report on the effects of three natural agonists on the contraction of smooth muscle in lung airways (Perez and Sanderson, 2005): (SMC = smooth muscle cell)

"Characteristics and Morphology of Lung Slices

"Throughout the lungs, the intrapulmonary airways and arteries have a close anatomical association that follows a parallel course. As a result, an airway and an accompanying arteriole (a bronchiole–arteriole pair) was easily identified and visualized in a single microscopic field of view (Fig. 1, A and B). This also makes the identification of pulmonary veins (not shown) easier because they are found as individual structures at some distance away from the bronchiole–arteriole pair. This anatomical separation precludes a direct comparison of the arteriole and vein responses in the same experiment. Each bronchiole–arteriole pair, when observed in transverse section, usually consists of a larger airway and a smaller arteriole. The airway is characterized by a lining of cuboidal epithelial cells with actively beating cilia. The arteriole lumen is lined with a low profile, squamous endothelium. Both structures are surrounded by a dense layer of tissue that often has a fibrous appearance. More distally, the airway and arteriole are surrounded by the alveolar parenchyma consisting of thin-walled sacs (Fig. 1, A and B). Specific antibody staining for SMC a-actin (Fig. 1 A) reveals that the SMCs are located in the surrounding fibrous layer, directly below the epithelium or endothelium. It is important to note that, in the lung slices used, only the alveoli remain filled with agarose. Before gelling, the agarose is flushed out of the airways with air. The gelatin is absent from the arterioles because it dissolves during incubation at 37°C. Consequently, the luminal compartments do not offer resistance to contraction. Agarose does not dissolve at 37°C but remains in the alveoli to keep the alveoli inflated and airways open."

Finally, here is an example of a complete *General Observations* subsection from my axon staining paper:

- "The normal Bodian stain recipe (listed in the *Materials and Methods*) works well on young amphibian tadpoles. As reported in the standard literature (e.g. Luna, 1960; Lillie, 1965), the Bodian stain highlights the cytoskeleton of neurons. In stained sections, mature axons and striated muscle appear black, while cell nuclei are dull brown or purple. The Bodian stain also enhances melanin pigment granules. In the embryonic central nervous system, the marginal zone remains unstained except for the locations of those few axons sufficiently mature to contain many neurofilaments. In Bodian-stained sections, the few large, early-maturing axons can be followed individually. An example is shown in Figure 1a."

3.3.2.   Subsection Two—*Specific Observations*

The middle subsection of your Results presents the heart of your research. Here you should report the complete data for your key variables. The *Specific Observations* subsection should also include the visual presentation—i.e., the table, chart, diagram, or graph—that you have already made to summarize your data. As a draft for the text summary of your data figure, use the descriptive paragraph you had written earlier and set aside.

3.3.2.1.   Narrative Descriptions and Pictures

Begin the *Specific Observations* subsection with those key variables that cannot be easily quantified.

Describe, for instance, the structural, textural, shape, or color details of your observations, and tell how these variables changed during the experiments. Look for patterns in the changes, and point them out to your reader. This is the part of the text that usually refers the reader to photographs, illustrations, and diagrams.

For example, in a paper on Guatemalan tarantulas, you might include this picture:

To illustrate the thoroughness needed in a good narrative scientific description, here are examples of nonnumerical observations reported in the *Results* sections of two scientific research papers:

- From a report on using the tri-block polymer P188 6h after a spinal cord injury in a mammal to increase the function and heal the structure of the spinal cord (Borgens et al., 2004):

"Comparative Study of Outcomes: 3D Anatomic Studies

"Due to the severity of the crush injury, injured segments of spinal cord were compressed and stenotic (Fig. 5i). For every 3D reconstruction, the injury site was centrally located at the region where the spinal cord was most compressed during registration of the images. The bilateral indentation of the crush injury was evident in many of the 3D reconstructions (Fig. 5Bi). Typically, after a constant displacement injury a hemorrhagic lesion is produced, causing profound destruction of the central gray matter of the spinal cord and variable sparing of the circumferential white matter (Moriarty et al., 1998; Tuszynskietal., 1999; Duerstock and Borgens, 2002). The 3D reconstruction processes both imaged and quantified regions of destroyed gray matter, marginal areas of spared gray and white matter, and cysts. Figure 5A and 5B include insets of histologic sections from the lesion epicenter of vehicle-treated and P188-treated reconstructed spinal cord segments, respectively. These histologic photomicrographs show thin tracts of spared silver impregnated axons of the white matter, whereas more central columns and most gray matter were destroyed. The images (ii) in Figure 5 show regions of apparently normal parenchyma in control (vehicle-treated) and P188-treated spinal cords. This intact parenchyma in control segments was located mainly in the subpial region and was virtually nonexistent at the lesion epicenter (Fig. 5Bii, Dii).

"Normal appearing parenchyma (both gray and white) was more abundant in P188-treated spinal segments than in control spinal cords (Fig. 5Aii, Cii). The amount of cavitation in these cords was the simplest pathologic feature to discriminate between the groups (Fig. 6). In the P188-treated spinal cord segments, cysts were smaller and dispersed more variably on either side of the central lesion (Fig. 6A, C). In the vehicle-treated spinal cords, cysts were larger and more localized within the spinal cord segment (Fig. 5, insets; Fig. 6B, D). Often cavitation was not confined within the field of view, extending beyond that particular segment of spinal cord (Fig. 5i). The area of scarification, devoid of silver impregnated nerve fibers, was located centrally at the region where the spinal cord was most compressed. In the vehicle-treated spinal cords, the lesion appeared less focal than in PEG-treated cords and tended to surround the cysts (Fig. 5, insets)."

- From a report describing the movements of neighboring cells in tissue culture (Abercrombie and Heaysman, 1954):

"The following is a brief qualitative description, based mainly on examination of films, of what happens between the two confronted explants in what we shall term the inter-explant area. The two sheets of cells, in a constantly changing irregular two-dimensional meshwork, looser in density at the periphery than centrally, advance towards each other, narrowing down the empty space between. Isolated cells may cross over from one sheet to the other. Contact is then established here and there between the peripheral cells of the two sheets and steadily becomes more general as the loose texture of each periphery is gradually lost, the intercellular spaces becoming smaller. So far, there is no noticeable diminution of the growth trend of the two sheets outward from their respective explants. But as the open spaces fill, concerted movement becomes more localised, except at the lateral margins of the inter-explant area, i.e., the sides not bordered by the explants. There some of the cells take part in a new trend towards the lateral unoccupied space. As the intercellular spaces become reduced to sizes considerably smaller than the area of a single cell, general trends of movement cease to be detectable, except laterally. The cells maintain only an uncoordinated oscillation. The continuous sheet of cells which now covers the inter-explant area is still largely a "monolayer," though with some overlapping of cell processes and here and there of entire cells."

And here is a good example of a qualitative, nonnumerical description that is based on numerical data:

• From a report on the effects of three natural agonists on the contraction of smooth muscle in lung airways (Perez and Sanderson, 2005): (ACH = acetylcholine, SMC = smooth muscle cell)

"To investigate the cause of the slow relaxation of the airway observed during the stimulation with ACH in the continued presence of Ni^{2+}, we simultaneously measured the changes in SMC $[Ca^{2+}]_i$ and contraction of the airway (Fig. 9C). After the initial increase in $[Ca^{2+}]_i$ and the onset of Ca^{2+} oscillations, the frequency of Ca^{2+} oscillations progressively decreased and this was accompanied by the relaxation of the airway. After removal of Ni^{2+}, but still in the presence of ACH, the frequency of the Ca^{2+} oscillations increased again and the airway recontracted. These results suggest that the Ni^{2+} blocked Ca^{2+} channels that contribute to the maintenance of the Ca^{2+} oscillations and perhaps serve to refill internal Ca^{2+} stores."

3.3.2.2.   Numerical Data

Next, present the numerical data about those key variables that you have quantified. The *Results* section should contain your experimental observations without much filtering or adjusting. Therefore, when your end-results are normalized (or otherwise adjusted) values, be sure to include at least some of the raw data values.

(a) A Few Numbers

If your experimental plan produced only a few data points, you can simply list the key variables for each of those data points. A table or chart might not be necessary.

For instance, in our hypothetical tarantula paper, we might write:

"We found only two tarantulas of the species *Brachypelma mythicosum* that had 6 dorsal stripes. These were a 10 cm long male and a 12 cm long female."

• Here is an example that is excerpted from the *Results* section of a research paper on the embryology of the peripheral nervous system. This paper reports on weeks of observations of the growth of individual axons and their accompanying sheath cells in living animals. The study produced only a few numbers from very long experiments (Speidel, 1932).

"III.   Rate of primitive sheath-cell migration

"Attempts were made to stimulate sheath cells to their maximum migratory possibilities by tail section and regeneration. Rapid movement was observed six days after the operation. Measurements showed that the more active cells in the newly regenerated zone sometimes moved more than 200μ during a twenty-four hour period. Many instances have been recorded of movement of from 100μ to 200μ. As yet, no case has been observed in which the movement exceeded 230μ."

(b) Many Numbers

On the other hand, if you have many data points, present your data organized into natural groupings in tables or graphs. Here is the place to put the visual data arrangement that you chose earlier during your exploratory analysis. Give the figure

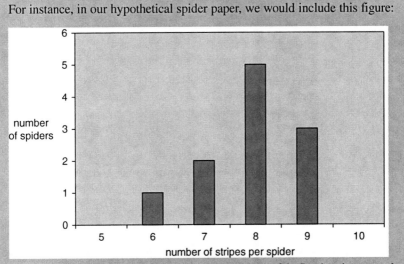

For instance, in our hypothetical spider paper, we would include this figure:

number of spiders

number of stripes per spider

Graph 1. Histogram of the number of dorsal stripes on each of the Brachypelma tarantulas collected near El Estor, Guatemala, during the summer of 2006. The sample included seven female and five male tarantulas, with body sizes ranging from 5–17 cm.

The narrative summary in the accompanying text might begin:

"Twelve tarantulas of the genus *Brachypelma* were photographed on days 1–3 of the expedition. Graph 1 shows the numbers of spiders with 6, 7, 8, or 9 stripes ..."

This summary would also include a statement about the patterns that we identified, e.g.:

"In our sample of 12 spiders, most (42%) had 8 stripes ..."

a self-explanatory legend, and describe the figure by inserting into the main text the summary statement you have already written.

(c) Examples of Tables

Here are examples of simple, clear tables from the *Results* sections of three scientific research papers:

• From a report on the role of juvenile hormone during the development of moths (reconstructed from Williams, 1961).

TABLE III
Effects of removal of corpora allata from fifth instar *Cecropia* larvae

		Results	
State at time of operation	No of animals	Normal pupa	Pupal-adult
Young 5th instar	3	0	3
Mid 5th instar	9	5	4
Late 5th instar	5	2	3
Spinning	13	12	1

• From a report on the relatively low number of studies on motor control in the psychological literature (reconstructed from Rosenbaum, 2005).

TABLE 1

Topic	Occurrence
Attention	51,946
Cognition	65,039
Decision or judgment or reasoning	54,367
Language	42,205
Memory	48,867
Motor	17,424
Perception or pattern recognition	34,328

Citations of selected topics in the social science citation index from 1986 to 2004. Note: All topics except for Motor are referred to in Ashcroft's (2002) Cognitive Psychology (3rd ed.; see Appendix B).

• From a report on using the tri-block polymer, P188, 6h after a spinal cord injury in a mammal to increase the function and heal the structure of the spinal cord (reconstructed from Borgens et al., 2004).

TABLE I.
Cutaneous trunchi muscle response to P188

Group	n	% Loss[a]	P[b]	Recovered[c]	P[b]	% Recovered[d]
P188	11	47.0 ± 2.6	0.13	5	0.03	40.8 ± 5.3
Control	10	41.3 ± 3.3		0		

[a] The percentage of the total cutaneous trunchi muscle (CTM) receptive field that was lost after spinal injury (mean ± SE).
[b] Statistical difference between the two groups compared using Fisher's exact test, two-tailed.
[c] The number of animals that recovered any portion of the CTM by 1 month post injury.
[d] The percentage of the area of areflexia that recovered after P188 treatment (mean ± SE).

(d) Examples of Graphs

Here are examples of clean, straightforward graphs from the *Results* sections of three scientific research papers:

- From a report describing the movements of neighboring cells in tissue culture (redrawn from Abercrombie and Heaysman, 1954).

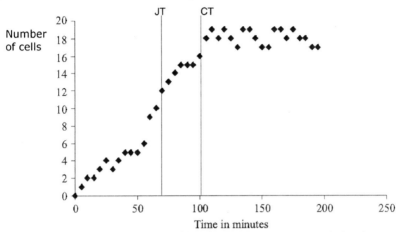

Figure 1. The number of cells in an area of interface between two explants during the approach and junction of the two sheets of cells, recorded at five-minute intervals. "JT" (Junction Time) is the time when opposing cells first make contact; "CT" (Close Fusion Time) indicates the time when the cells have become closely packed throughout the area observed.

- From a report on the unmitigated cerebral suppression of sound from the ipsilateral ear in split-brain patients (redrawn from Milner et al., 1968).

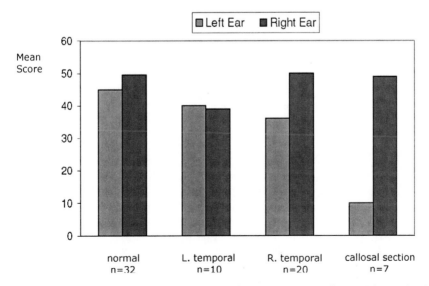

Figure 1. Mean number of digits correctly reported for each ear, when different digits are simultaneously presented to the two ears. The results are for normal control subjects and for three different patient groups.

• From a report on the aging changes of Au/n-AlGaN Schottky diodes when exposed to air (redrawn from Readinger and Mohney, 2005).

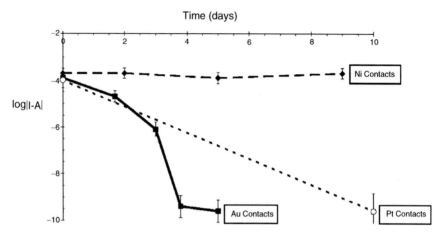

Figure 4. The J(rev) at –5V for Au, Pt, and Ni diodes on n-AlGaN following exposure to air.

3.3.3.   Subsection Three—*Case Studies*

Finally, present the full, unretouched details of one or two specific examples from your data set. Your reader will understand your experiments better after seeing raw observations with all the complexities of the real world.

How do you choose these raw examples? There are two equally acceptable categories: best cases and representative cases.

3.3.3.1.   Best Cases
Best cases show an extreme. They are the biggest, littlest, roundest, slowest, or fastest. Best cases fall at one end of the spectrum of your actual results—they show what *can* happen, not necessarily what usually happens.

For instance:

• If you are studying stripes on the backs of tarantulas, pick one of the spiders that had the most or the least stripes. You might write, "Only 2 of the tarantulas we found had <7 stripes. Tarantula C4 was one of those ... " Then include a drawing or photograph of tarantula C4 with its dorsal stripes clearly visible.

(continued)

(continued)

> Write a brief note describing where and when that particular tarantula was found, its gender, its size, its color, and any other distinguishing features.
> - If you are studying cell staining, you may find that your new stain is finicky. Perhaps the stain works infrequently, but when it does, the results are dramatic. Although good examples are not common in your experiments, you can report these cases as best-case examples. You might write, "Our new stain worked unpredictably, but we could usually find sections of the slide where cells stained crisply and in great detail. A good example is shown in Figure 1b. Two mature axons can be seen coursing longitudinally in parallel in the ventral marginal layer (large arrows). In addition, five tiny immature axons can clearly be seen growing radially, hugging the cell layer (small arrows)."

- Here is an example of a best-case description from a report on the growth of individual axons and their accompanying sheath cells, as followed in the living animal over periods of days to weeks (Speidel, 1932).

"One of the most striking cases of primitive-sheath-cell activity and function is shown in the series of figure 11. A survey of the fins of this tadpole on May 30th revealed a place where a primitive sheath cell (M) appeared to be in the act of transferring from a larger nerve (ramus lateralis vagi, dorsal branch) to a very slender delicate fiber (fig. 11, a). This fiber (PP) extended distally for some distance beyond the field and was entirely devoid of primitive sheath cells. It could also be traced proximally for several myelin segments, as a non-myelin-emergent fiber. The nearest primitive sheath cell on this fiber was four myelin segments away. The appearance of the cell (M) suggested ameboid activity. For four days, however, its position remained unchanged, although the contour of the nucleus varied somewhat. It seemed that the region, instead of being one of active change, was rather one of stability. The transfer of this cell, however, was completed on June 4th (fig. 11, b). Cell division took place and the proximal daughter cell (M1) moved toward the terminal myelin segment (fig. 11, c). This cell transferred next to the myelin-emergent nerve sprout (fig. 11, d), and divided, giving rise to two cells (M1a and M1b, fig. 11, e). Thickening of the sprouts took place under the influence of these cells, and on June 9th two new myelin units were visible (fig. 11, f). On June 11th, these had increased both in diameter and in length (fig. 11, g). Likewise, the sprouts emerging from them became more complex. The distal primitive sheath cell (M2, fig. 11, d) also divided (fig. 11, e), but each of the daughter cells (M2a and M2b) remained quiescent and exerted no influence in stimulating myelin formation.

"This is in marked contrast to the effect of the two cells which transferred to the myelin-emergent sprouts. ..."

3.3.3.2. Representative Cases

Alternatively, you can show representative cases. Representative cases are modal cases, the most common examples in your data set.

> For instance:
> - If you are studying stripes on the backs of tarantulas, pick one of the spiders that had the most common number of stripes. You might write, "Forty-two percent of the tarantulas we found had 8 dorsal stripes, the most common number. A good example was Tarantula B1 …" Include a drawing or photograph of tarantula B1 with its dorsal stripes clearly visible. Write a brief note describing where and when that particular tarantula was found, its gender, its size, its color, and any other distinguishing features.
> - If you are studying cell staining, you may find that your new stain usually works well. You might write, "Figure 1b shows an example of the detail that is usually visible after using our stain. Two mature axons can be seen coursing longitudinally in parallel in the ventral marginal layer (large arrows). In addition, five tiny immature axons can clearly be seen growing radially, hugging the cell layer (small arrows)."

- Here is an example of a representative case description from a report on the role of juvenile hormone during the development of moths (Williams, 1961).

"[O]nly two of the eight animals transformed into normal pupae. The other six formed strange creatures in which a considerable number of tissues and organs had overleaped the pupal stage by undergoing precocious adult differentiation.

"One of these animals is illustrated in Figure 2. The head shows the pigmented, faceted, compound eyes of the adult. The antennae exhibit the segmentation and subdivisions characteristic of early adult development. The thorax is covered with a mixture of rugose pupal cuticle and smooth cuticle of the adult type. The adult patagia have developed at the base of the fore-wings. The sclerotization of the thoracic tergum is adultoid. The thoracic pleura and sternum are covered for the most part by a smooth, adult-type cuticle. The legs show segmentation and the differentiation of tarsal claws and pulvilli. The proximal ends of the wings are covered by adult cuticle. The cuticle of the abdomen is wholly pupal, except in the immediate region of the genitalia; the latter are represented, not by imaginal discs, but by miniature adult genitalia which show an early elaboration of the various valves and adult structures. Dissection revealed that the fat-body was similar to that of a pupa after the initiation of adult development. Moreover, the ovaries showed the differentiation of ovarioles to a stage corresponding to that encountered in early adult development."

3.4. An Example of a Complete *Results* Section

"Results

The normal Bodian stain recipe (listed in the *Materials and Methods*) works well on young amphibian tadpoles. As reported in the standard literature (e.g. Luna, 1960; Lillie, 1965), the Bodian stain highlights the cytoskeleton of neurons. In stained sections, mature axons and striated muscle appear black, while cell nuclei are dull brown or purple. The Bodian stain also enhances melanin pigment granules. In the embryonic central nervous system, the marginal zone remains unstained except for the locations of those few axons sufficiently mature to contain many neurofilaments. In Bodian-stained sections, the few large, early-maturing axons can be followed individually. An example is shown in Figure 1a.

Our 10–12 µm thick tissue sections were quite sensitive to the intensifier. Intensification for more than 3 min led to heavy black staining of cytoskeletons and of melanin granules, with a general black staining of the background. In contrast, intensification for 1 min produced only light background staining while clearly enhancing the cytoskeletons of cells. At 1 min intensification, the procedure did not stain structures (such as mitochondria) that are not normally blackened by the Bodian stain. Rather, 1 min intensification further darkened those cell structures that were only faintly stained by normal Bodian stain techniques.

Figure 1b shows an example of the detail that was usually visible after using the intensifier. Two mature axons can be seen coursing longitudinally in parallelin the ventral marginal layer (large arrows). In addition, five tiny immature axons can clearly be seen growing radially, hugging the cell layer (small arrows)."

"Cell cultures

Fixed frog and chick cell cultures stain poorly with the standard Bodian techniques. Nuclei are the most prominently stained structures, taking on a dull brown or purple color. With Bodian staining, the thin cell processes, such as lamellipodia and axons, are almost invisible under standard microscopic illumination and require phase-contrast or DIC (Nomarski) imaging to be faintly seen. Examples are shown in Figures 1c and 1e.

On the other hand, our intensification procedure made cell processes readily visible under standard microscopic illumination. Examples are shown in Figures 1d and 1f. All cell types were stained clearly, and the thinnest cell process could be seen using DIC imaging. For example, microspikes could be clearly distinguished along the axon and at the axonal growth cone. Figure 2 shows one example of the most detailed and most clearly stained growth cones. Optimal enhancement required 10 min of post-staining intensification for cell cultures, as opposed to the 1 min required for thick tissue sections."

4.   DISCUSSION

> ### *Skeleton of the* Discussion
>
> A. Recap Your *Recipe* → *Results* Report
> B. Archive Your Results
> C. If Possible, Make a Proposal

The archives of science are enormous, and without some effort, your data will be lost among the millions of other scientific observations already in storage. A well-written *Discussion* section will help to ensure that your results will be both visible and accessible in these archives. To accomplish this, your *Discussion* should do two things:

- First, it should present a clear, concise summary of your data.
- Second, it should link your observations to those of other scientists in one or both of these ways:
  - ○ Include an annotated list comparing specific aspects of your data to data that is already in the scientific archives.
  - ○ Use your data and related data from the scientific archives to generate a proposal, a generality, a theory, or a model.

4.1.   Recap Your Results

Begin your *Discussion* with a recap, a short summary of what you learned about the key variables of your data.

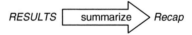

Here is an example of the *Recap* subsection, from a mythical paper on tarantulas:

"During three consecutive days of collecting in the summer of 2006, we caught, measured, and photographed 12 tarantulas of the species *Brachypelma mythicosum* in the jungles of Guatemala. Seven of these tarantulas were females and five were males. The median body length was 11.5 cm (range = 5–17 cm). All 12 spiders had characteristic white dorsal stripes. The modal number of dorsal stripes was 8 (range = 6–9)."

Here are examples of *Recap* paragraphs that begin the *Discussion* sections in a variety of research papers.

- A report on the Kaposi's sarcoma-associated herpesvirus (KSHV) effects on infected human cells (Glaunsinger and Ganem, 2004):

"We demonstrated previously that lytic KSHV replication promotes a widespread shutoff of cellular gene expression that likely occurs via enhanced mRNA turnover. In this report, we have further characterized the consequences of this mRNA shutoff using microarray-based expression profiling. Our findings reveal that very few human transcripts accumulate during infection, including most of those induced by vGPCR."

• A report of measurements of the durations of brief localized calcium currents ("concentration microdomains") in a presynaptic axon terminal (Sugimori et al., 1994):

"To understand in more detail the nature of these concentration microdomains, we obtained rapid video images (4,000/s) after injecting the photoprotein n-aequorin-J into the presynaptic terminals of squid giant synapses. Using that experimental approach, we determined that microdomains evoked by presynaptic spike activation had a duration of about 800 ps. Spontaneous quantum emission domains (QEDs) observed at about the same locations as the microdomains were smaller in amplitude, shorter in duration, and less frequent. These results illustrate the time course of the calcium concentration profiles responsible for transmitter release. Their extremely short duration compares closely with that of calcium current flow during a presynaptic action potential and indicates that, as theorized in the past, intracellular calcium concentration at the active zone remains high only for the duration of transmembrane calcium flow."

• A report on using high doses of cytosine arabinoside (ARA-C) to keep patients with acute myeloid leukemia (AML) in continuous complete remission (CCR) (Bohm et al., 2005):

"Repetitive cycles of high dose ARA-C (HiDAC; ARA-C, $3\,g/m^2$ twice a day on days 1, 3, and 5) have been introduced as an effective postremission treatment for patients with AML by the CALGB study group. However, only a few confirmatory reports on the effects of this protocol have been published so far. We here report the results of HiDAC consolidation for AML obtained in a single center. The overall percent survivals of the remitters, the DFS, and the CCR at 4 years were 41, 34, and 39%, respectively. Our results are in line with results of the CALGB study group that reported an overall survival of remitters of 46%, a DFS of 39%, and a CCR of 44% at 4 years in their patients. Together, these data demonstrate that repetitive cycles of HiDAC represent a highly effective consolidation strategy for patients with AML in CR."

• A report on using the tri-block polymer P188 6h after a spinal cord injury in a mammal to increase the function and heal the structure of the spinal cord (Borgens et al., 2004):

"The use of tri-block copolymers in reversing membrane permeabilization in other injury and ischemia models were introduced earlier. We add to this emerging picture a beneficial usage of tri-block copolymers as an acute treatment for neurotrauma. In particular, we report that a single subcutaneous injection of P188 6hr after a standardized compression injury to the adult guinea pig spinal cord produced these three major findings:

o the recovery of SSEP conduction was enhanced by P188 treatment (90% of the population) compared to an insignificant spontaneous recovery in the control group;
o the midthoracic injury produced a similar area of areflexia in the CTM reflex in both groups, although a statistically significant recovery of these silent receptive fields occurred only in response to P188 treatment;
o 3D reconstruction of all serial histologic sections comprising the injured segments of spinal cord showed that P188 treatment reduced the amount of pathologic cavitation of the cords and was associated with an increased volume of intact parenchyma."

• A report on the effect on animal diets of S-methylmethionine (SMM), a compound found only in plants (Augspurger et al., 2005): (Met = methionine)

"The objective of these experiments was to determine the bioavailability of SMM as either a Met or a choline source, using a Met-and choline-deficient semi-purified diet previously shown to respond markedly to dietary additions of either nutrient. The marked responses in growth performance when SMM was added to choline-deficient, Met-adequate diets in Assays 1, 3, and 4 showed that SMM exhibited choline-sparing bioactivity; indeed, SMM exhibited 20% bioequivalence (wt:wt) to choline per se. Conversely, scrutiny of the data in Assays 1 and 2 suggested that SMM probably did not exhibit Met activity in chicks."

• A report examining the ultrasound echogenicity of the substantia nigra in Parkinson's disease (Berg et al., 2001): (PD = Parkinson's disease, SN = substantia nigra)

"This study confirms our previous findings of increased SN-echogenicity in PD patients. The proportion of distinctly hyperechogenic SNs in the group of PD patients, however, was higher which is likely to reflect improvements in the ultrasound technology in the last five years. In 91% of PD patients, the extent of hyperechogenic signals at the SN was well beyond the standard deviation of an age matched control group while only 9 out of 103 PD patients (8.7%) exhibited SN signal intensity within the normal range. Moreover, PD patients with a more extended hyperechogenic signal had an earlier disease onset. Additionally, these PD patients more often showed motor complications like fluctuations, dyskinesia, and freezing when analysis was controlled for the duration of disease. ..."

4.2.   Archive Your Results

After cleanly summarizing your results, help to archive them by showing the connections between your research and research reported by other scientists.

Begin by making two tables of related scientific papers. One table should compare your results to results of papers reporting experiments that used *recipes* similar to yours.

Similar recipes		Results
Your recipe	→	{ a b c d e f h }
Paper A's recipe	→	{ b d e f g }
Paper B's recipe	→	{ a b d f }
Paper C's recipe	→	{ b d e h }

This table lists research reports that used the same general experimental techniques as you used, and it compares the results that both you and they discovered.

The other table should compare your recipe to the recipes used in papers that reported *results* similar to yours.

Similar results		Recipe
{ a b c d e f h }	←	Your recipe
{ a b c d e f h }	←	Paper D's recipe
{ a b c d e f g h }	←	Paper E's recipe
{ b c d e f g h }	←	Paper F's recipe

This table lists research reports that attempted to study the same things as you studied but that used different experimental techniques from yours.

Now, translate your table(s) into a clean narrative text for your *Discussion*.

To illustrate how these two types of comparisons have been written in a wide variety of *Discussions*, here are some examples.

(a) Comparisons of results among papers using similar recipes, i.e., descriptions of what other scientists found using similar experimental approaches:

- From a report of measurements of the durations of brief localized calcium currents ("concentration microdomains") in a presynaptic axon terminal (Sugimori et al., 1994):

"The time course reported here is different from that observed in other secretory systems such as the chromaffin cell (14) where the transient develops over 1 to 100 ms and where the calcium buffering properties may be quite different. However, our results are similar to the calcium concentration profile in individual frog-muscle sarcomeres following the activation of action potentials (14)."

- From a report of growth of individual axons and accompanying sheath cells followed in the living animal over periods of days to weeks (Speidel, 1932):

"Sheath cells in general migrate from proximal to distal. During early development, such centrifugal movement is especially noticeable. It is also obvious in actively generating areas in which the growth urge is strong (compare figs. 1, 2, and 13). However, in the later stages after the first flush of growth is over, the primitive sheath cells now occasionally migrate in a proximal direction (centripetal movement). This is particularly noticeable in connection with the transfer of primitive sheath cells to myelin-emergent sprouts (compare figs. 8 to 11). Harrison's ('24) statement that he has never seen sheath cells migrate proximally is probably to be explained by the fact that his observations were on very young tadpoles in which the centrifugal growth urge was still quite strong."

- From a report on the comparison of the atomic weight of ordinary lead with lead of radioactive origin (Richards and Lembert, 1914):

"This matter has received not only speculative but also experimental treatment at Harvard. For many years, the possibility that samples of a given element from different sources might have different atomic weights had been considered, and investigated, but never before with a positive outcome. In the first investigation of the atomic weight of copper undertaken by one of us as long ago as 1887,[7] samples of copper obtained from Germany and from Lake Superior were found to give precisely the same atomic weight for this element. More recently, the question was revived and in 1897, specimens of calcium carbonate were obtained from Vermont, U. S. A., and from Italy, in order to discover whether the calcium in these two widely separated localities had the same atomic weight. Not the slightest difference was found between them.[8] Again, in a very elaborate investigation on the atomic weight of sodium,[9] silver was obtained partly from several distinct sources and sodium chloride was obtained partly from several different samples of German rock salt, and partly from the salt pumped from the Solvay Process Company's mines at Syracuse, N. Y. These preparations, differing widely in the steps of manufacture and in geographical source, all yielded essentially the same atomic weights within the limit of error of the process.[10] Yet more recently Baxter and Thorvaldson, [11] with the same possibility in mind, determined the atomic weight of extraterrestrial iron from the Cumpas meteorite, which gave a result identical with ordinary iron within the limit of error of experimentation. From these researches, it would seem probable that even if an unusual eccentricity may be exhibited by lead, most elements do not as a rule differ from any such cause of uncertainty. Baxter and Grover are now engaged in the examination of ordinary lead from different geographical sources. Perhaps this also contains more than one component, as suggested above."

- From a report on the use of vital dye to determine the fates of parts of the amphibian neural plate (Jacobson, 1959):

"The external limit of the brain-forming area
"The limit between presumptive brain and presumptive skin has previously been dealt with in a number of papers based upon vital staining. Woerdeman (1929) and Fautrez (1942) were of the opinion that the boundary coincides with the outer limit of the neural plate. Already in 1939 it had, however, been clearly shown by Baker & Graves that the inner half of the transverse ridge takes part in the formation of the brain. The same result was reached by Horstadius and Sellman (1946) who, on the basis of their own results and those of earlier authors, discussed the problem in detail. Their result differs from those of the present investigation in that they indicate (in their Fig. 5c, p. 15) that material belonging to the first four of the zones distinguished by them in the neural ridge participates in the formation of the brain. According to my division, these zones would closely correspond to the ridge level with I and half of II (Text-fig. 1). Experiments VC 26, VC 27, VC 38, VC 51, VD 7, and VD 11 show that the limit of the presumptive brain cannot be situated upon the ridge farther back than the region which corresponds to Horstadius & Sellman's zone 1 plus the cranial part of zone 2. Experiment VD 24 and others show that this limit, where it is situated upon the transverse ridge, must be drawn as in Text-fig. 16."

- From a report on the use of proton magnetic resonance spectroscopy to assess the severity of damage in a shaken infant (Haseler et al., 1997):

"Numerous MRS studies are now available to document the metabolic changes in infant brain subject to a variety of injuries. To our knowledge, we are the first to indicate the nature and severity of the biochemical response to traumatic injury, but Auld et al., (16) describe other patterns in a variety of nontraumatic (17,18) brain injuries. Hypoxic injury also results in loss of NAA and accumulation of lactate. The pattern seen here in SBS appears somewhat different from hypoxic injury alone (19). A larger study is required to establish the specificity, if any, of any of the differences seen in our small series, to SBS."

- From a report on using high doses of cytosine arabinoside (ARA-C) to keep patients with acute myeloid leukemia (AML) in continuous complete remission (CCR) (Bohm et al., 2005): (CALBG = Cancer and Leukemia Group B, HiDAC = high dose ARA-C)

"Another prognostic parameter examined was age. It is generally accepted, that younger patients (aged <60 years) have a superior outcome compared to elderly patients [27, 31, 32]. The data of the CALGB study group have also demonstrated that age is a predictive factor for the outcome of therapy in younger AML patients [16]. These results could be confirmed in the present study. In fact, the outcome of therapy in patients aged under 40 years was significantly better compared to the survival in older patients (age 40–60 years). All in all, HiDAC seems to be a most effective postremission regimen in younger patients with favorable karyotype, but also in a group of patients with intermediate karyotype. In this regard it should be stated that our patients in general may resemble a relatively good risk group of AML (<60 years, CR). ...

"So far, it remains unknown whether the intensity of the postremission treatment can be reduced. Notably, only a few data exist indicating how many consolidation cycles are required to maintain long term CR in patients with AML [28,29]. In the original HiDAC protocol, a maximum of four cycles of consolidation were applied. Thus, the primary intention of the present study was to give a total number of four consolidation cycles as well. Indeed, a slight majority (56.8%) of the 44 patients undergoing consolidation (at least one cycle of HiDAC) received all four planned consolidation cycles. Another 22.7% of these patients received three cycles of HiDAC. It is of interest, that in the group of patients receiving three cycles of HiDAC, we were able to identify several patients who showed long-term disease-free survival. Therefore, it is tempting to speculate that a reduction of the number of HiDAC consolidation cycles to three may be possible and may be equally effective compared to four cycles [28]."

(b) Comparisons of recipes among papers studying similar results, i.e., descriptions of what other scientists found when looking at the same endpoints but using different experimental approaches:

• From a report on the effect on animal diets of S-methylmethionine (SMM), a compound found only in plants (Augspurger et al., 2005): (Met = methionine)

"The presence of SMM in foods and feeds adds one more complicating factor to the proper assessment of the requirement and bioavailability of choline. In mammalian species, but not avians, excess dietary Met can eliminate the dietary need for preformed choline (17, 18, 36, 37), and betaine, which is also present in many foods and feeds (38), can replace the methylation function of choline (31). Thus, previous attempts to obtain clear-cut animal responses to choline supplementation of corn-soybean meal diets have often met with failure (18, 39), probably due not only to significant concentrations of choline per se in soybean meal but also to the presence of both betaine and SMM in this ingredient. Moreover, previous research in which the bioavailability of choline per se in soybean meal was found to be as high as 80 (22) to 100% (40) probably represents over estimates. It is apparent that when considering a dietary need for bioavailable choline, choline itself as well as betaine and SMM levels in foods and feeds must be taken into account. For mammalian species, including humans, excess dietary Met is also capable of sparing choline."

• From a report on the relatively low number of studies on motor control in the psychological literature (Rosenbaum, 2005):

"Another reason to expect motor control to become more popular in psychology is the emergence of ecological psychology and dynamical systems analysis. Advocates of ecological psychology argue that the primary function of perception is to guide action (Gibson, 1979) and that the control of action enlists rather than resists physical properties of actor-environment couplings (Bernstein, 1967). Advocates of dynamical systems analysis seek to describe ongoing cycles of perceiving and acting in the form of differential equations (e.g., Sternad, Duarte, Katsumata, & Schaal, 2001). The advent of the ecological and dynamical systems perspectives has fostered the analysis of classes of behavior that were left out of the research portfolio of traditional cognitive psychological research, which focused on internal representations and computations to the exclusion of embodied cognition (Clark, 1997; Glenberg, 1997). Newly studied topics include walking and jumping (Goldfield, Kay, & Warren, 1993; Thelen, 1995), juggling (Beek & Turvey, 1992), skiing (Vereijken, Whiting, & Beek, 1992), pistol shooting (Arutyunyan et al., 1969), wielding objects (Carello & Turvey, 2004), bouncing a ball on a tennis racquet (Sternad et al., 2001), swinging two handheld pendulums of different lengths and weights (Turvey, 1990), and oscillating two index fingers at different frequencies and relative phases (Zanone & Kelso, 1997)."

- From report on the effects of three natural agonists on the contraction of smooth muscle in lung airways (Perez and Sanderson, 2005): (ACH = acetylcholine, SMCs = smooth muscle cells)

"Previous investigations of Ca^{2+} signaling in rat (Tolloczko et al., 1995, 1997) or dog (Yang et al., 1997; Yang, 1998) tracheal SMCs used low-speed sampling systems to report that 5-HT induced an initial transient followed by a sustained elevation in $[Ca^{2+}]_i$. By contrast, we found with video-rate confocal microscopy that 5-HT induces repetitive transients in $[Ca^{2+}]_i$ or Ca^{2+} oscillations in airway SMCs. In most respects, the 5-HT–induced Ca^{2+} oscillations were similar to those induced by ACH, both in this and previous studies with lung slices (Bergner and Sanderson, 2002a, 2003) or isolated tracheal airway preparations (Prakash et al., 1997, 2000; Roux et al., 1997; Kuo et al., 2003). The Ca^{2+} oscillations persisted with a steady frequency and usually originated at one end of the cell and spread toward the other end as a Ca^{2+} wave, although the direction of the Ca^{2+} waves could be reversed. The Ca^{2+} waves were unsynchronized between neighboring cells and did not propagate to adjacent cells, suggesting that each wave originated within each cell. It is important to note that each Ca^{2+} oscillation did not generate a twitch of contraction but that the SMCs maintained a steady contractile state."

4.3.   Try to Make a Proposal

Your *Discussion* should always recap your data and it should help to archive the data (i.e., it should show notable connections between your *recipe → results* report and other existing scientific papers.) Beyond this, your *Discussion* can sometimes offer a proposal, that is, a conjecture, prediction, generalization, hypothesis, model, or theory.

To illustrate how specific proposals have been written in a wide variety of *Discussions*, here are some examples.

- From a report consolidating evidence of the speed of extinction of North American dinosaurs (Fastovsky and Sheehan, 2005):

[Proposal: *A geologically instantaneous event – an asteroid impact – caused the mass extinction of the dinosaurs in North America.*]

"The conclusion that the extinction of the dinosaurs was geologically instantaneous precludes longer-term causes (e.g., events on million to ten-million-year timescales). So, although survivorship patterns may be in accord with habitat fragmentation-based models, habitat fragmentation as the driving force for the dinosaur extinction is problematical, because it is linked in this case to a marine regression that occurred over a million or more years. Moreover, recent stratigraphic work summarized in Johnson et al. (2002) suggests that the Hell Creek was deposited rather quickly (over ~1.4 m.y.; Hicks et al., 1999, 2002) in a transgressive setting (the final transgression of the North American Western Interior Sea). This interpretation is concordant with a previously inferred rise in the water table (Fastovsky and McSweeney, 1987). The transgressive geological setting is antithetical to the proposed fluvial lengthening associated with the habitat fragmentation scenario and suggests that it was not likely a factor in the North American dinosaur extinction.

"**Death by Asteroid**

"The current "alternative hypothesis" for the cause of the extinction of the dinosaurs is, of course, an asteroid impact with Earth. Schultz and d'Hondt (1996), using the morphology of the crater as an indicator of the angle and direction of the impact, proposed that the Western Interior of North America would bear the brunt of impact effects. In all scenarios, wholesale extinctions on extremely short timescales are presumed to be a consequence of such an event. While the extinction cannot be shown to have occurred within hours, days, or weeks, extinction timescales can be constrained to a few tens of thousands of years or less. For this reason, what is known of the rate of the dinosaur extinction in North America is concordant with the predicted effects of an asteroid."

- From a report examining the ultrasound echogenicity of the substantia nigra in Parkinson's disease (Berg et al., 2001): (PD = Parkinson's disease, SN = substantia nigra)

[Proposal: *The substantia nigra nuclei in brains of patients with Parkinson's disease will have higher than normal concentrations of iron.*]

"The reason for the increase in echogenicity is still unclear. Morphological changes occurring at the SN in PD which may lead to an alteration in tissue impedance and therefore SN echogenicity are several: Loss of pigmented neurons may result in tissue condensation, proliferation of microglia may increase cellular interfaces, and elevated heavy metal tissue content (especially an increase of iron) [9, 13, 16, 27, 28] may modify tissue impedance. According to recent findings iron is supposed to be a major source for the increased echogenicity of the SN; neurochemical and sonographic analyses of post mortem material revealed a close correlation between SN echogenicity and iron tissue content (Berg et al., unpublished). In addition, animal experiments have demonstrated more intense tissue echogenicity induced by increasing amounts of iron injected into the SN [4].

"Therefore, we surmise that SN hyperechogenicity in PD might (at least in part) be due to disease-associated elevation of SN iron levels. Iron is supposed to play a pivotal role in the degeneration of SN neurons in PD, as it facilitates and accelerates neurodegeneration by formation of free radicals and lipid peroxidation [14, 20, 30]. In the light of these observations, one may speculate that increased echogenicity of the SN might reflect higher tissue iron content which, on the other hand, could increase the oxidative stress within the SN resulting in a more rapid degeneration of SN neurons."

- From the *Discussion* in a report of growth of individual axons and accompanying sheath cells followed in the living animal over periods of days to weeks (Speidel, 1932):

[Proposal: *Three specific behaviors will be seen in most myelinating nerves.*]

"I propose, in a tentative way, the following new principles of neurogenesis for the peripheral nerves to supplement the list given by Cajal. All of these relate to the mechanism of nerve fiber myelination.
1. A myelin-emergent nerve sprout differs from a non-myelin-emergent fiber. The former in combination with a primitive sheath cell leads to the formation of a new myelin segment, the latter ordinarily does not.
2. The transfer of a primitive sheath cell from a non-myelin-emergent fiber to a myelin-emergent fiber may be effected in a variety of ways, but the reverse transfer is rare. Transfer from one myelin-emergent sprout to another may also take place.
3. Early unmyelinated nerves serve, in a general way, to direct advancing myelin-emergent nerve sprouts, and to furnish them with primitive sheath cells as a preliminary step of myelination ..."

- From a report describing the movements of neighboring cells in tissue culture (Abercrombie and Heaysman, 1954):

[Proposal: *Fibroblasts have an innate resistance to crawling over each other.*]

"Our results lead to the conclusion that there is some restriction on the movement of fibroblasts over each other's surfaces. It is a restriction that does not appear to operate at a distance to hinder the mutual approach of fibroblasts. Fibroblasts therefore freely make contact with each other and, in fact, adhere together (5) forming the meshwork so characteristic of their growth. The restriction operates only after contact has been established. There is evidence (1) that the initial reaction to contact is a slight acceleration of the movement of the cells towards each other. After that, there usually occurs a prohibition of further movement in this direction. The prohibition does not invariably occur in the conditions we have investigated, since there develops some overlapping of cells; and we have as yet no conclusive information as to how important the prohibition is in cultures grown for more that 36 hours, or grown in different media.

"This directional prohibition of movement we shall refer to briefly as 'contact – inhibition'."

- From a report on the apparent disintegration of nitrogen atoms by radioactive bombardment (Rutherford, 1919):

[Proposal: *Controlled beams of alpha particles could be used to break apart nuclei of light atoms.*]

"Taking into account the great energy of motion of the alpha particle expelled from radium C, the close collision of such an alpha particle with a light atom seems to be the most likely agency to promote the disruption of the latter; for the forces on the nuclei arising from such collisions appear to be greater than can be produced by any other agency at present available. Considering the enormous intensity of the force brought into play, it is not so much a matter of surprise that the nitrogen atom should suffer disintegration as that the alpha particle itself escapes disruption into its constituents. The results as a whole suggest that, if alpha particles–or similar projectiles–of still greater energy were available for experiment, we might expect to break down the nucleus structure of many of the lighter atoms."

- From a report of improvements to a model of the release kinetics of solute from a polymeric matrix, with experimental tests of this improved model (Paul and McSpadden, 1976):

[Proposal: *The new mathematical model will accurately describe the rate of solute release from polymeric matrices in most real-world situations.*]

"We conclude that the refinements of the Higuchi model offered here (via the relaxation of the 'pseudosteady-state' assumption) have some advantages for describing release kinetics for loadings where $A/C(s)$ is slightly greater than one, but become virtually identical with Higuchi's equation for large values of $A/C(s)$. The incorporation of a finite external mass transfer resistance into the models for release kinetics makes these results more valuable for describing situations normally encountered in most applications. The refined model in conjunction with the classical solution to Fick's law for A/C(s) < 1 offer an accurate set of equations for describing release rates over the entire range of A/C(s), provided of course the condition of rapid dissolution of undissolved solute is applicable."

- From a report of measurements of the durations of brief localized calcium currents ("concentration microdomains") in a presynaptic axon terminal (Sugimori et al., 1994):

[Proposal: *Approximately 15,000 calcium ions flow into a presynaptic axon terminal during an evoked microdomain.*]

"The number of calcium ions that are detected in an evoked microdomain may be estimated as follows. Consider that a presynaptic action potential generates an I(Ca) of about 300 nA (10, 13), a current of 0.5 pA/channel, and that a single channel allows the flow of about 150 to 200 calcium ions (7, 18), then ~6 × 10^5 channels are opened to release 5,000 to 10,000 vesicles (4, 11) or ~15,000 calcium ions/vesicle (13). [This is in contrast to he estimated minimum of 200 ions/vesicle (one calcium channel) for the ciliary ganglion (18)]. Given that the number of active zones in a presynaptic terminal is about 5,000 to 10,000 (6), the number of channels open per action potential for each active zone is ~100. If single channels allow the flow of 150 to 200 calcium ions (7, 18), an evoked microdomain, using n-aequorin-J and the present imaging technique, may represent an influx of calcium of about 15,000 ions."

- From a report on the use of proton magnetic resonance spectroscopy to assess the severity of brain damage in a shaken infant (Haseler et al., 1997):

[Proposal: *Much of the damage to the brain of a shaken infant is secondary to the release of destructive enzymes from injured cells.*]

"Lysosomal damage with the release of acid hydrolases,[9] and perhaps most importantly, phospholipase A_2[8] is recognized as a cause or at least an accompaniment of neuronal damage. It is possible that deleterious effects on neurons are more marked in those immature neurons in which final axonal connections are incomplete. The speed with which the loss of neuronal marker NAA occurs ($t_{1/2}$ ~2 days) in the only infant from whom accurate data are available (B), and from the earliest data point available in infant C, suggests an accelerated process triggered by an initial injury (shaking). Autocatalytic enzymatic processes are one of a number of potential mechanisms to describe such an event. This cascade may explain why the degree of neurological damage suffered is greater than the amount of trauma apparently involved. We postulate that trauma itself, although not the sole cause of neuronal injury, is the initiating event that may release enzymes that cause the continued development of neurological damage after trauma has ceased."

• From a report on the Kaposi's sarcoma-associated herpesvirus (KSHV) effects on infected human cells (Glaunsinger and Ganem, 2004):

[Proposal: *Cellular (native) molecules promoting continued viral replication are among the handful of proteins still being produced 10–12 hours after KSHV lytic infection of a spindle cell.*]

"It is interesting that three of the induced genes in Table II are involved in cellular transcription. In particular, HIF-1α is a heterodimeric basic helix-loop-helix transcription factor that becomes stabilized under hypoxic conditions. Stabilized HIF-1α dimerizes with HIF-1β and transcriptionally activates several genes responsive to low oxygen whose products play critical roles in tumor progression including angiogenesis, cell growth, and energy metabolism (24). Although we did not observe the escape of many HIF-1 target genes in infected cells, one HIF-1 target is up-regulated; that is, the transcript for solute carrier family 2, member 3 (GLUT3). Interestingly, it has been reported that hypoxia also induces KSHV lytic reactivation (25), an event likely mediated by the presence of functional hypoxia response elements within the RTA and ORF34 promoters. Thus, an additional function for HIF-1 in lytic cells may be to enhance KSHV replication via the stimulation of select viral promoters, the products of which are not subject to shutoff."

• From a report on the effects of three natural agonists on the contraction of smooth muscle in lung airways (Perez and Sanderson, 2005): (CICR = calcium-induced calcium release, RyR = ryanodine receptor channel, SMCs = smooth muscle cells, SR = sarcoplasmic reticulum)

[Proposal: *Calcium ion oscillations in smooth muscle cells involve the repeated alternate overload and release of intracellular calcium ions stored in the sarcoplasmic reticulum.*]

"From this data, we hypothesize that the KCl-induced Ca^{2+} oscillations are the result of the following events. Initially, KCl induces membrane depolarization and initiates an influx of Ca^{2+} via L-type and/or T-type Ca^{2+} channels. The cell compensates for this rise in $[Ca^{2+}]_i$ by transporting the extra Ca^{2+} into the SR via SERCA pumps. Because of their limited capacity, the stores quickly overload as indicated by the increasing frequency of the elemental Ca^{2+} events, which reflect sensitized RyRs. Upon reaching a critical Ca^{2+} load, the elemental Ca^{2+} events trigger an extended phase of CICR via sensitized RyR to empty the store and generate a Ca^{2+} wave with a transient SMC contraction before initiating the cycle again."

- From a report on the role of juvenile hormone during the development of moths (Williams, 1961):

[Proposal: *In insects, juvenile hormone acts to stop the decoding of the genetic information that leads to continued developmental differentiation.*]

"From this summary, we learn that the role of juvenile hormone is to modify the cellular reactions to ecdyson. It appears to do so by opposing progressive differentiation without interfering with growth and molting in an unchanging state. In some unknown manner, it blocks the de-repression and de-coding of fresh genetic 'information' without interfering with the acting-out of information already at the disposal of cells."

- One of the most famous biological proposals was made in the *Discussions* of a number of papers by James Watson and Francis Crick for the replication of the genetic information of a cell. Here is a quotation from one of those *Discussions*:

[Proposal: *The molecular replication of genetic information includes the splitting of a DNA helix into two complementary chains, and the recreation of the original double-chained helix from each of the separated chains.*]

"Previous discussions of self-duplication have usually involved the concept of a template, or mould. Either the template was supposed to copy itself directly or it was to produce a 'negative', which in its turn was to act as a template and produce the original 'positive' once again. In no case has it been explained in detail how it would do this in terms of atoms and molecules.

"Now our model for deoxyribonucleic acid is, in effect, a pair of templates, each of which is complementary to the other. We imagine that prior to duplication, the hydrogen bonds are broken, and the two chains unwind and separate. Each chain then acts as a template for the formation on to itself of a new companion chain, so that eventually we shall have *two* pairs of chains, where we only had one before. Moreover, the sequence of the pairs of bases will have been duplicated exactly."

- [From Watson JD, Crick FHC (1953) Genetical implications of the structure of deoxyribonucleic acid. *Nature* 171(4361): 964–967]

4.4.   An Example of a Complete *Discussion* Section

> "**Discussion**
>
>   We have described an intensification procedure for enhancing the Bodian staining of fixed cell structures. The intensification highlights and defines more clearly those cytoskeletal structures normally made only faintly black by Bodian silver. The Bodian silver stain has a high affinity for neurofilaments (Gambetti et al., 1981; Phillips et al., 1983). Our intensification procedure clearly reveals neurofilament bundles that are barely stained by standard Bodian techniques.
>
>   Isolated neurons show their innate characteristics best in vitro. In tissue culture, the movements and the interactions of neurites can be followed in microenvironments that can be controlled precisely (Harrison, 1910; Speidel, 1932, 1933; Pomerat et al., 1967; Tennyson, 1970; Bunge, 1976; Pfenninger and Reese, 1976; Bray, 1982; Letourneau, 1985). Our intensification procedure offers a new tool for tissue culture studies of embryonic and immature axons, both of which have relatively few neurofilaments.
>
>   For tissue culture, our Bodian stain intensification has these four features:
> a) it works for standard tissue culture procedures
> b) it works on a variety of tissues
> c) the degree of intensification is approximately proportionate to the duration of treatment
> d) it effectively reveals cell processes as thin as microspikes in fixed light microscopic preparations."

5.   CONCLUSION

> *Skeleton of the* Conclusion
>
> One paragraph statement of the point of the paper

Each research paper should present only one or two main ideas, and these ideas should be stated in the *Conclusion*. The *Introduction* of the paper should show the current need for these ideas, the *Discussion* should tie these ideas to other existing scientific papers, and the *Conclusion* should summarize the ideas in one succinct paragraph. Some journals use a format that includes a section labeled "Conclusion" or "Summary." For other journals, the *Conclusion* is the untitled last paragraph of the *Discussion*.

When you first face the *Conclusion* section of your paper, you should already have a draft of the *Discussion*. The *Discussion* moves from your specific observations to more general statements relating your data to the work of others. To write a *Conclusion*, take the recap from the beginning of your *Discussion* and the general statements from the remainder of your *Discussion*, and forge a single uncluttered paragraph.

Discussion:

For example, the *Conclusion* section of my axon staining paper was:

- **"Conclusion.** Our intensification of the standard Bodian stain successfully outlines fine cell processes of neurons, both in tissue sections and in fixed tissue cultures. The details of individual growth cones of cultured neurons are especially clearly stained and can then be easily seen with light microscopy."

Here, from a variety of scientific articles, are examples of straightforward *Conclusions* that clearly state the points of the papers:

- From a report describing the movements of neighboring cells in tissue culture (Abercrombie and Heaysman, 1954):

"It is concluded that fibroblasts avoid moving over each other's surfaces. Such 'contact-inhibition' of movement can explain why it is that fibroblasts normally migrate predominantly radially from an explant and that the whole culture tends rapidly to become circular in plan whatever its initial form."

- From a report on using high doses of cytosine arabinoside (HiDAC) to keep patients with acute myeloid leukemia (AML) in continuous complete remission (CCR) (Bohm et al., 2005):

"Together, our data provide further evidence that post-remission therapy with HiDAC is a safe and effective consolidation treatment for AML patients in CR aged less than 60 years. The optimal number of consolidation cycles and the subgroups of patients who benefit most from this regimen remain to be defined in forthcoming trials."

- From a report on using the tri-block polymer P188 6h after a spinal cord injury in a mammal to increase the function and heal the structure of the spinal cord (Borgens et al., 2004): (PEG = polyethylene glycol, ROS = free radicals, LPO = lipid peroxidation)

"In summary, P188 may be a free radical scavenger (Marks et al., 2001), whereas PEG is not. They both, however, directly reduce ROS and LPO in the damaged nervous system. Furthermore, they both provide neuro-protection to injured spinal cord and thus will continue to be investigated as potential therapies, simple to apply, for various forms of neurotrauma."

- From a report consolidating evidence of the speed of extinction of North American dinosaurs (Fastovsky and Sheehan, 2005): (K-T = Cretaceous-Tertiary period boundary)

"In the 25 years since Alvarez et al. (1980) first proposed that an impact was responsible for the K-T extinctions, stratigraphic and paleoecologic evidence have come together to present a reasonably cohesive picture of a quick demise of the dinosaurs. Evidence from the rates of dinosaur extinction suggests that the extinction was geologically instantaneous; this conclusion in combination with the nature of the post-Cretaceous biologic recovery suggests that the extinction occurred on an extremely short, irresolvable time scale. While the exact killing mechanisms may or may not yet have been identified, all the data—including the rate of extinction, the nature of the recovery, and the patterns of survivorship—are concordant with the hypothesis of extinction by asteroid impact."

- From a report on the unmitigated cerebral suppression of sound from the ipsilateral ear in split-brain patients (Milner et al., 1968):

"The fact that all the commissurotomized patients were able to report digits presented to the left ear without difficulty, when there was no competing input from the right ear, shows that the ipsilateral pathway could be utilized. The suppression of ipsilateral input in the presence of a competing stimulus from the contralateral ear provides clear behavioral evidence of the dominance of the contralateral auditory projection system in man, a finding for which there is by now considerable electrophysiological and some behavioral support from work with lower species."

- From a report on the effects of three natural agonists on the contraction of smooth muscle in lung airways (Perez and Sanderson, 2005): (SMC = smooth muscle cell)

"In conclusion, intrapulmonary airways respond to 5-HT and ACH with a contraction that is maintained by high frequency Ca^{2+} oscillations within the SMCs that arise from repetitive cycles of Ca^{2+} release and uptake by the SR and require extracellular Ca^{2+} for store refilling. By contrast, KCl-induced twitching of SMCs results from low frequency Ca^{2+} oscillations produced by an overfilling and uncontrolled release of internal Ca^{2+}. Most importantly, the magnitude of the contraction of airway SMCs is regulated by the frequency of the Ca^{2+} oscillations."

- From a report examining the effect of matrix metalloproteinase inhibitors on healing after periodontal surgery (Gapski et al., 2004): (BOP = bleeding on probing, ICTP = marker for level of bone resorption, LDD = low dose doxycycline, PD = probing depth)

"Six-month administration of LDD suggests that there is an enhanced postsurgical wound healing compared to placebo controls with regard to PD reduction. This positive effect was most marked in deep sites (≥7mm), where the differences in PD reduction were maintained until the completion of the trial. Reductions in the bone resorption marker ICTP were also found in patients while on the drug, suggesting the potential of LDD to act as a bone-sparing agent. In addition, the percentage of BOP sites was affected by LDD therapy, but this effect was only noticeable during the period of the drug administration. Finally, no significant shifts on the periodontal microbiota beyond that attributable to the effects of the surgery could be seen with the utilization of LDD."

- From a report examining the ultrasound echogenicity of the substantia nigra in Parkinson's disease (Berg et al., 2001): (PD = Parkinson's disease, SN = substantia nigra, TCS = transcranial ultrasound)

"Our study demonstrates that TCS may serve as a valuable tool in the neuroimaging of PD providing easily available information in addition to other neuroimaging data. Because of the lack of invasiveness and the relatively low cost, it is particularly useful for an application to a large number of patients. Further studies are required to determine whether differences in the echogenicity of the SN in PD patients may display differences in the genetic background or other patho-genetical factors of the disease."

- From a report showing the weight of radioactive lead isotopes to be different from the weight of ordinary lead (Richards and Lembert, 1914):

"The outcome [of our experiments] was striking. There can be no question that the radioactive samples [of lead] contain another element having an atomic weight so much lower than that of ordinary lead as to admit of no explanation through analytical error, and yet so nearly like ordinary lead as not to have been separated from it by any of the rather elaborate processes to which we had subjected the various samples."

- From a report on the cellular processes that bring about healing of fingertip injury in monkeys (Singer et al., 1987):

"We conclude that repair of amputated distal digits does not occur in the same manner as in lower vertebrates. Our results confirm earlier reports in the human (see references in the *Introduction*) that the conservative open wound management of patients with such injuries is a satisfactory method that usually produces a cosmetically attractive, normally sensitive, and useful digit."

6.   LIMITATIONS OF THIS STUDY

Skeleton of the Limitations

A. Qualifier no. 1
B. Qualifier no. 2
C. ...

No experiment is perfectly unambiguous, and your experimental observations will always come with caveats, assumptions, and limitations. For instance, you may not have been able to randomize your experimental subjects, the perfect analytic machine may have been too expensive for your budget, technology may not yet be sufficiently advanced to simultaneously measure all the variables you needed, you may have been forced to end your experiments after 3 years rather than 4 because of uncontrollable outside reasons, or your statistical analyses may be based on the unprovable assumption that your data come from a normally distributed set of values.

The results of real world experiments often need qualifiers. In your paper, your *Materials and Methods* section should detail the limits of your techniques, your *Results* section should describe the range and variation of your data, and your *Discussion* section should state clearly the assumptions you have used when formulating your general statements and proposals.

Explicitly stated qualifiers help the reader to evaluate the strength of the details throughout your paper. It is equally important to summarize the major qualifiers of your experiments alongside your conclusions. This can be done in a brief section, titled "Limitations of this Study," placed immediately after the *Conclusions* section.

In the *Limitations of this Study*, write the most significant qualifiers and provisos, each as a single, brief paragraph. The paragraph should:

• Describe the activity that had the limitation—i.e., an experimental technique, an analytic technique (e.g., a statistical test), or a line of reasoning.
• Explain the limitation.
• Suggest how the limitation may have affected one of your conclusions.

Here is a good example of a *Limitations of this Study* section from a paper that explores the effect that hypothermia has on the length of time (length of stay, 'LOS') a patient needs to stay in the recovery room (postanesthesia care unit, 'PACU') after surgery (Kiekkas et al., 2005).

"**Study Limitations.** The present study was not a randomized, controlled trial; thus, baseline patient and procedure characteristics were not equally distributed between hypothermic and normothermic patients. Mean duration of surgery was significantly different between the two groups.

"Hypothermic patients tend to shiver more during the postoperative period than normothermic patients; consequently, they more often are treated with IV opioids, increasing the time required for observation in the PACU. The difference in appropriate LOS may be a result of not only hypothermia per se, but also of the fact that the groups of hypothermic and normothermic patients were not homogeneous.

"The study population was restricted to patients undergoing orthopedic surgery, and the mean age of this population was rather high. Whether the present findings can be generalized to other categories of surgical patients, therefore, needs to be determined."

7.   INTRODUCTION

> *Skeleton of the* Introduction
>
> A. Background
>   1. Currently-Accepted General Statements
>   2. Available Supporting Data
> B. Gap
> C. Your Plan of Attack

In its *Materials and Methods* and *Results* sections, a research paper describes a set of recipes and the data that those recipes produced. In its *Discussion* section, the paper attempts to fit this data into the overall database of science, which is a many-dimensional, continually evolving jigsaw puzzle. The task of the *Introduction* is to give the reader a preview of the *Discussion*, pointing out in advance the particular hole in the scientific landscape that the paper's data will try to fill.

The *Introduction* section begins by orienting the reader. It describes a part of the scientific puzzle that is complete, a region with well-recognized landmarks and clearly defined contours. The *Introduction* then leads the reader down a short, direct path toward an empty space—a gap in the puzzle—and announces, "Our data should fit here."

Beginning of *Discussion*:

"Here is a well known landmark"

End of *Discussion*:

"My experiments should fit here."

7.1.   Define the Gap

Your research paper will report a set of *recipe* → *results* observations and it will then organize the observations in a form that can be summarized in one or two sentences. These summary sentences will be your *Conclusion*.

The *Introduction* section of your paper should set the stage for your *Conclusion*. Specifically, the *Introduction* should describe the gap in our current scientific knowledge that can be filled by your *Conclusion*.

By the time you begin to write your *Introduction*, you should already know your *Conclusion*, and therefore the specific gap that your paper will fill can be described simply by rephrasing the summary statements from your *Conclusion* as questions. If the *Conclusion* of your study is "256 angels can dance on the head of a pin," then the gap in our current knowledge would be the answer to, "How many angels can dance on the head of a pin?" Your *Introduction* should raise this question and then point out that it cannot be answered by currently available scientific data.

> For example, suppose that, in our fictitious paper about the stripe patterns on the backs of tarantulas, the *Conclusion* is "The number of dorsal stripes on Guatemalan tarantulas ranges between 6 and 9, with the most commonly observed number (the mode) being 8 stripes." Knowing this, we would write our *Introduction* so that it explains that, at the moment, researchers cannot answer the question, "How many dorsal stripes can we expect to find on Guatemalan tarantulas?"

7.2.   Begin with the Known

To make the argument that there is currently a hole in our scientific knowledge, begin your *Introduction* at a firm place in the scientific archives, that is, start with a scientific statement that is widely accepted. Then lead your reader step-by-step from the known to the unknown, the gap that your *Conclusion* will fill.

When choosing where to start, think about the audience of the journals for which you are writing. Pick a starting point that most of its readers should already know or accept.

> For my report on nerve staining, for example, I planned to send the final paper to a journal read by a wide variety of histologists. Some of these readers might not be familiar with the relatively specialized Bodian stain, and others might not know about the difficulties of staining embryonic axons. Therefore, I began my *Introduction* more broadly, with a historical statement that should have been part of the education of most histologists, "Silver staining of neurons began in the 19th century, when Camillo Golgi found that nerve cells have a strong affinity for silver salts."

7.3. Take a Direct Path to the Unknown

From this spot of firm ground, take your reader into the specific area of your research problem by following a short chain of previously reported observations. Lead the reader straight to the place where your *Conclusion* should be, and pose the question that your *Conclusion* answers. Explain that the answer is currently unknown, and show your reader the edges of this hole in our knowledge by citing the closest information available in the scientific literature.

> In the *Introduction* to my axon staining paper, I went from Golgi's invention of the silver stain to Cajal's use of the stain in an encyclopedic axon-mapping project. I explained that after Cajal, Bodian developed a simpler silver stain that was especially good for tracing axon tracts. Finally, I pointed to the current gap in our technology: as useful as the Bodian stain has proved for highlighting many different varieties of neurons, the stain does not work well on immature or tissue culture axons.

In your *Introduction*, provide sufficient references so that readers can go to the scientific literature and see for themselves the particular observations that currently surround the hole you propose to fill.

Your Conclusion should fill this gap.

7.4. Summarize Your Plan-of-Attack

After leading your reader to a gap in our knowledge, end your *Introduction* by stating briefly how you plan to fill the gap. Your plan-of-attack comprises the recipes detailed in your *Materials and Methods*. Therefore, the last few sentences of your *Introduction* should summarize the main recipes that have given you the data on which your *Conclusion* is based.

Your plan-of-attack should be a variant of the statement, "Here we report the observations that can be seen after doing X," where 'X' is a summary of the key recipes in your *Materials and Methods*.

The scientific gap

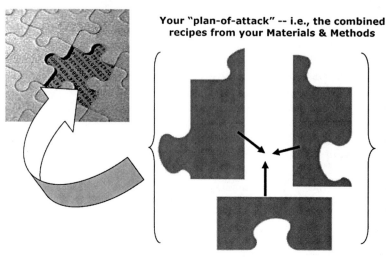

Your "plan-of-attack" -- i.e., the combined recipes from your Materials & Methods

At the end of the *Introduction*, briefly tell us how your experiments should end up filling an existing gap in the scientific archives.

7.5.   A Sampler of *Introductions*

Scientific papers are written from diverse perspectives and about a wide range of subjects. Although all scientific papers should be built using the same basic principles, the particular field and the type of data of each research project shape the papers individually. Here are examples of how a number of scientific writers have used the same basic skeleton

> A. Background
>   1. Currently-Accepted General Statements
>   2. Available Supporting Data
> B. Gap
> C. Your Plan of Attack

to take the reader from a currently-accepted general statement to their proposed plan-of-attack for a variety of quite different research projects.

• From a report on the role of juvenile hormone during the development of moths (Williams, 1961):

A. Background
1. Currently-Accepted General Statements

"In the history of every hormone, there is a more or less prolonged period when the factor is recognizable only as a certain 'activity' within a living system. Progress at this stage is largely dependent on the development of a method of biological assay that is simple, selective, and quantitative.

"In the case of the juvenile hormone of insects, the pioneering studies that led to the discovery of the hormone simultaneously directed attention to a method for its assay. This method, as developed by Wigglesworth (1936, 1948, 1958), is performed on mature nymphs of *Rhodnius* ..."

2. Available Supporting Data

"Numerous investigators, following Wigglesworth's lead, have utilized the 'larval assay' in testing the endocrine activity of corpora allata. The literature includes studies of the following genera: *Bombyx* (Bounhiol, 1938; Fukuda, 1944; Ichikawa and Kaji, 1950), *Dixippus* (Pflugfelder, 1939, 1958), *Tenebrio* (Radtke, 1942), *Galleria* (Piepho, 1942, 1950b), *Melanoplus* (Pfeiffer, 1945), *Drosophila* (Vogt, 1946), *Gryllus* (Poisson and Sellier, 1947), *Oncopeltus* (Novak, 1951), and *Calliphora* (Possompes, 1953). The conclusions derived in all these studies have confirmed the fact that the corpora allata undergo substantial changes in activity during the course of metamorphosis."

C. Gap

"At the Harvard laboratory, we also have tried to make use of the larval assay in testing for juvenile hormone. A survey of all our experiments during the past fifteen years fails to reveal a single instance in which the implantation of active corpora allata interfered with the transformation of fifth stage *Cecropia* larvae into normal pupae. For reasons that are not fully understood, the larval assay does not work in the case of the *Cecropia* silkworm."

D. The Plan of Attack

"Solution of our problem came from an unexpected direction. As described in the previous paper of this series, the 'pupal assay' was accidentally discovered in 1947; unlike the larval assay, it proved to be an extremely sensitive test for juvenile hormone (Williams, 1952a, 1959).

"In the present study, the pupal assay has been used as a principal tool in a study of the role of juvenile hormone in the growth and metamorphosis of the *Cecropia* silkworm."

- From a report of growth of individual axons and accompanying sheath cells followed in the living animal over periods of days to weeks (Speidel, 1932):

A. Background
1. Currently-Accepted General Statements

" While many studies have been made on the growth, development, and regeneration of peripheral nerves, the process of formation of the myelin sheath has never been adequately described."

2. Available Supporting Data

"Nerve cells from spinal ganglion, spinal cord, and brain have been cultivated in vitro, and the growth of processes from these has been watched (Harrison, '08, '10; Ingebrigsten, '13; Esaki, '29, and others). Likewise, the movements of the sheath cells of Schwann have been seen."

B. Gap

"No one, however, up to the present time has succeeded in observing the formation of the myelin sheath, either in tissue culture or in a living animal."

C. The Plan of Attack

"Studies of phenomena of degeneration and regeneration in frog tadpoles for the past few years (Speidel, '29) convinced me that it might be feasible to watch the activities of peripheral nerves in the tail-fin expansion of living tadpoles. Preliminary observations showed that the technical difficulties were not too great and that, under favorable conditions, it was possible to keep under observation from day to day the same sheath cells and nerve sprouts for a considerable period of time. This investigation, therefore, was undertaken in an attempt to follow individual sheath cells and nerve sprouts and to correlate their movements with the formation of the myelin sheath."

- From a report on the effect on animal diets of S-methylmethionine (SMM), a compound found only in plants (Augspurger et al., 2005): (Met = methionine)

A. Background
1. Currently-Accepted General Statements

"S-Methylmethionine (SMM) is an analog of S-adenosyl-methionine (SAM; Fig. 1), with a methyl group substituted for the adenosyl group. This compound is unique to plants and is found in measurable-to-high concentrations in corn, cabbage, tomatoes, celery, spinach, and garlic (1–3) ..."

2. Available Supporting Data

"Previous research elucidated the mechanism for synthetic synthesis and the in vitro and in vivo metabolism of SMM (7–12). SMM promoted the growth of *Escherichia coli* heterotrophs only in the presence of Cys or Hcy (9). Radiolabeling experiments suggested that the methyl group of SMM is available for choline or creatine synthesis, but not for Met synthesis (13)."

B. Gap

"In vivo experiments with rats were equivocal, with SMM supporting growth of rats fed sulfur amino acid (SAA)-deficient diets (14, 15), but not in rats fed SAA-free diets (13). Additionally, there is anecdotal evidence for SMM exhibiting choline bioactivity as a result of ameliorating choline deficiency–induced fatty liver in rats (16)."

C. The Plan of Attack

"The objective of this research was to qualitatively and quantitatively determine the efficacy of L-SMM for sparing dietary choline or Met. Avians, unlike mammals, have a requirement for preformed choline that cannot be replaced by excess Met (17, 18), making them an excellent animal model for studying choline bioavailability."

- From a report on the effects of three natural agonists on the contraction of smooth muscle in lung airways (Perez and Sanderson, 2005):

A. Background
1. Currently-Accepted General Statements

"Gas exchange in the lungs requires an appropriate matching of ventilation to blood perfusion and this is influenced by the caliber of the airways and blood vessels. Consequently, an understanding of the mechanisms that control the size of airways and arterioles is required to understand lung physiology and the development of obstructive lung disease and pulmonary hypertension."

2. Available Supporting Data

"In general, the mechanisms that control the caliber of intrapulmonary airways or arterioles have been investigated in either whole lungs or isolated smooth muscle cells (SMCs)."

B. Gap

"While these approaches provide valuable data, it is difficult to determine the site or size of airway or arteriole contraction by measurements of air flow in whole lungs or blood pressure in the pulmonary artery. Similarly, it is difficult to relate changes in intracellular Ca^{2+} concentration ($[Ca^{2+}]_i$) of isolated SMCs to the contractile responses of intact airways and arterioles."

C. The Plan of Attack

"Our solution to investigate how cellular physiology regulates the contraction of the small airways or arterioles was to examine living lung slices that retain many structural and functional properties of the lung. Relatively thick lung slices have been used to study the contractile response of airways (Dandurand et al., 1993;

Martin et al., 1996; Minshall et al., 1997; Adler et al., 1998; Duguet et al., 2001; Martin et al., 2001; Wohlsen et al., 2001), but thinner lung slices, combined with confocal microscopy, provided us with the ability to study changes in $[Ca^{2+}]_i$ of SMCs that underlie airway contraction (Bergner and Sanderson, 2002a, b, 2003)."

- From a report on using high doses of cytosine arabinoside (ARA-C) to keep patients with acute myeloid leukemia (AML) in continuous complete remission (CCR) (Bohm et al., 2005):

A. Background
1. Currently-Accepted General Statements

"In response to induction chemotherapy with cytarabine and an anthracy-cline, the majority (70–80%) of all patients with de novo acute myeloid leukemia (AML) aged less than 60 years enter complete hematologic remission (CR) [1–4]. However, without further postremission treatment, recurrence of disease is likely to occur. Thus, it is generally appreciated that postremission therapy is important to maintain CR in patients with AML [1–5]. Using 'standard dose chemotherapy' for consolidation, only 25% of these patients achieve a long lasting (continuous) complete remission [6, 7]. Therefore, alternative strategies of postremission therapy have been proposed. A straightforward approach is allogeneic stem cell transplanta-tion from a sibling donor [8, 9]. However, this procedure is restricted to younger patients with a suitable donor. For those patients who cannot be transplanted, con-solidation protocols employing high doses of ARA-C (monotherapy or in combina-tion with other cytostatic drugs) or autologous stem cell transplantation are usually considered as appropriate therapy [10–15].

"In 1994, the Cancer and Leukemia Group B (CALGB) published a high dose ARA-C-based consolidation regimen. This regimen (HiDAC) consists of repeti-tive (up to four) cycles of high dose ARA-C ($3\,g/m^2$) given twice on days 1, 3, and 5 [16]. In this particular CALGB study trial, HiDAC was found to represent an effective consolidation. Thus, the rates of leukemia-free survival were similar to those achieved with high dose chemotherapy and consecutive autologous stem cell transplantation [16]. In addition, the HiDAC protocol was reported as a relatively safe approach with moderate side effects in patients aged less than 60 years, and a low rate of treatment-related deaths."

2. Available Supporting Data

"Despite the apparent efficiency and relatively low rate of side effects, only a few studies have confirmed the value of high dose cytarabine as postremission treatment in patients with AML so far [14, 17, 18].

B. Gap

However, these studies did not follow exactly the protocol published by the CALGB.

> ### C. The Plan of Attack

We here present the outcome of 44 patients with AML treated with up to four cycles of HiDAC in a single center."

- From a report measuring the interface tensions in immiscible fluid mixtures as the fluid boundaries disappear (Sundar and Widom, 1987):

> ### A. Background
> ### 1. Currently-Accepted General Statements

"One of the striking features of the approach to a critical point of the equilibrium of two fluid phases is the disappearance of the interface between them and the vanishing of the associated interfacial tension. In the case of an ordinary critical point, for example that of the liquid-vapor equilibrium of a simple fluid or the liquid-liquid equilibrium of a partially miscible binary mixture, the interfacial tension s is known to vanish on approach to the critical point as (ref. 1)

$$s \approx (T_c - T)\mu$$

where T is the temperature and T_c, the critical temperature, with $\mu \approx 1.26$.

"The work described here addresses the corresponding question of the behavior of the interfacial tensions, and, in particular, the value of the exponent μ, on approaching a tricritical point of the equilibrium of three liquid phases in a quaternary system. ..."

> ### 2. Available Supporting Data

"The expectation from theory is that $s \approx (T_t - T)^\mu$ with $\mu = 2$ and T_t the tricritical temperature. This result follows from the van der Waals theory of the interface (ref. 4), with the free energy in the three-phase region near the tricritical point as given by Griffiths (ref. 5). This value of μ is thus expected to differ from the $\mu \approx 1.26$ associated with an ordinary critical point of two-phase equilibrium.

"The prediction of $\mu = 2$ was independently made (ref. 6) for the tricritical point that occurs in mixtures of ^{3}He and ^{4}He and has been verified experimentally there (ref. 7). This problem has also been studied in the context of the tricritical point that occurs in quasibinary mixtures of ethane + (n-octadecane + n-nonadecane) (ref. 8). There, an exponent that from theory is expected to be $(2/3)\mu$ was found to vary from about 1.26 to 1.4 for different paths of approach to the tricritical point. That would correspond to a variation of μ in the range 1.9 to 2.1. While a value in that range is in fairly good agreement with the theoretical value of 2, the variability is not understood."

> ### B. Gap

"While no experiments have been reported on systems of four or more components with the aim of determining μ, the results of surface tension measurements on brine + hydrocarbon + surfactant systems with a series of surfactants, at a single temperature, have been combined and the results, if suitably interpreted, were shown again to be consistent with $\mu=2$ (ref. 9)."

C. The Plan of Attack

"The system chosen for study here was the quaternary mixture ammonium sulfate + water + ethanol + benzene. The phase diagram for this system in the region of three-liquid-phase equilibrium near the tricritical point has been studied in detail (ref. 2) and the tricritical temperature determined to be 49°C.

"At each of a few different temperatures, the interfacial tensions $s_{a,b}$ and $s_{b,c}$ were determined for several overall compositions. These were then used to obtain the quantities $s_o(T)$ and $s_{(a, b), c}(T)$ defined above."

- From a report on the aging changes of Au/n-AlGaN Schottky diodes when exposed to air (Readinger and Mohney, 2005):

Background
Currently-Accepted General Statements

"The variable Al-Ga composition of n-type $Al_{(x)}Ga_{(1-x)}N$ provides new device structures and device capabilities unobtainable with n-GaN alone. High-quality Schottky barriers to n-type AlGaN are important for the efficient operation of ultraviolet photodetectors or heterostructure field-effect transistors (HFETs) (1–4)."

2. Available Supporting Data

"Several publications report the barrier height for Schottky contacts to n-AlGaN (5–13), including three studies of the influence of the Al mole fraction on the barrier height (8, 11, 12) where the increase in Al mole fraction ($x < 0.35$) provides higher Schottky barriers over those of pure n-GaN. Other more recent publications report on the thermal stability of Schottky barriers for gate metallizations on HFETs (11, 14–22). All of these studies were carried out on $n-Al_{(x)}Ga_{(1-x)}N$ with $0 \leq x \leq 0.35$, with a wide range of barrier heights reported (0.45–1.55 eV), similar to the case for n-GaN."

B. Gap

"We had originally intended to only study the electrical characteristics of Schottky contacts to n-AlGaN as prepared using different wet-chemical pre-metallization surface treatments. However, we found instead that another variable much more dramatically influenced the performance of Schottky contacts to n-AlGaN (23). In these studies, it was found that the measured Schottky barrier characteristics were not stable with time when exposed to air at room temperature. In fact, we observed a significant change in the electrical characteristics as barrier heights increased, ideality factors decreased, and reverse leakage currents were reduced after only hours or days."

C. The Plan of Attack

"In this paper, we summarize our previous work and present new studies of the environmental aging of Schottky barriers. A summary of our findings as well as the proposed mechanism shall also be presented."

- From a report on the cellular processes that bring about healing of fingertip injuries in monkeys (Singer et al., 1987):

A. Background 1. Currently-Accepted General Statements

"The ability of lower vertebrates, such as fish, newts or salamanders, to regenerate body parts missing either by accident or by deliberate alteration in the laboratory is well known (Wallace 1981; Geraudie and Singer 1984)."

2. Available Supporting Data

"These animals can replace an amputated extremity (limb or fin) with a new distal structure of normal morphology and function. Two factors have been shown to be of great importance in the initiation of the regrowth of the body parts in these lower forms.

"The first factor is the necessity for a fresh wound surface, which soon will be covered by a pluri-stratified epidermal epithelium originating from the margins of the amputation site (Repesh and Oberpriller, 1978; Stocum, 1985). Suturing of the amputation site as well as repeated removal of the epidermal regenerating cover prevents regrowth (Thornton, 1957, 1968; Singer, 1980).

"The second requirement involves the presence of an adequate nerve supply at the wound surface met either by motor or sensory axons (Singer, 1946, 1952, 1974). The regeneration process after epidermal wound healing consists of an accumulation of cells, the origin of which has excited controversy in the past (Wallace, 1981). Nevertheless, recent work using a monoclonal antibody to an intracellular antigen specific for a subpopulation of blastemal cells supports the hypothesis that blastemal cells could arise by a process of dedifferentiation of the stump tissues, especially muscles and Schwann cells, as shown by Brockes's study (1984). The accumulation phase is followed by the differentiation phase when cells begin to differentiate and reconstruct the limb tissues: cartilage giving way to endochondral bone, muscles, and connective tissue invaded by blood vessels and nerves emerging from the stump."

B. Gap

"The regenerative capacity of lower vertebrates to replace perfectly an amputated extremity was thought to be nonexistent in higher vertebrates, including man. In recent years, however, the open wound treatment of fingertip injuries has given very satisfactory results according to numerous authors (Bojsen-Moller et al., 1961;

Holm and Zachariae, 1974; Illingworth, 1974; Bosley, 1975; Farrell et al., 1977; Fox et al., 1977; Rosenthal et al., 1979; Louis et al., 1980; Chow and Ho, 1982) which has also been the experience of one of the authors (ECW) in a small number of cases (unpublished results). In fact, the repair has been so good especially in small children that the question arises whether this might actually be regeneration similar to that just described for lower forms of life. In fact, in mammals, mice have been shown to regrow the extremity of an amputated toe (Neufeld, 1980; Borgens, 1982). Consequently, in order to study this problem, we decided to test it on the *Rhesus* monkey, a mammal closer to humans."

C. The Plan of Attack

"In this paper, we report our results of the repair process following digital amputation of digits in the above animal. The data deals with the study of the morphology and histology of regrowth and quantification of innervation in amputated digit tips studied with histological silver nerve impregnation and counting of cross-sectioned axons."

7.6.   An Example of a Complete *Introduction* Section

"Introduction

Silver staining of neurons began in the 19th century, when Camillo Golgi found that nerve cells have a strong affinity for silver salts. By 1880, he had modified Louis Daguerre's 1839 recipes for developing silver iodide photographs and had created a silver stain for fixed neural tissue. Golgi's silver stain cleanly highlighted the full three-dimensional arborization of the axon and the dendrites of individual neurons. With Golgi's stain, Santiago Ramon y Cajal (1928) mapped the cellular architecture of a wide variety of nervous systems. His comprehensive silver stain studies remain the foundation of neuroanatomy (Santini, 1975; Parent, 1996).

In the 1930's, David Bodian developed a silver staining recipe that was easier and more consistent than earlier methods (Bodian, 1936). The Bodian stain has a high affinity for neurofilament proteins (Gambetti et al., 1981; Phillips et al., 1983), and Bodian staining can be used to trace individual axons through thick tissue sections with a light microscope (e.g. Katz and Lasek, 1981). Loots et al. (1979) and Rager et al. (1979) review many of the useful modifications of Bodian's original stain recipe.

Embryonic and immature axons have thin neurofilament bundles. None of the Bodian staining techniques blacken small bundles of neurofilaments sufficiently to be resolved cleanly with light microscopy. For this reason, the Bodian silver stains have not been useful for studies of the earliest stages of nervous system development or the fine details of axons growing in tissue culture. Here, we report on the ability of a post-staining intensifier to enhance Bodian staining of young axons and axons in tissue culture."

---

8.   ABSTRACT

> *Skeleton of the simple* Abstract
>
> One Paragraph: "We did. We saw. We concluded."
>
> – or –
>
> *Skeleton of the* Abstract *with subsections*
>
> A. One or Two Sentences "BACKGROUND"
> B. Two or Three Sentences "METHODS"
> C. Less Than Ten Sentences "RESULTS"
> D. One Sentence "CONCLUSION"

The *Abstract* presents the essence of your *Materials and Methods*, your *Results*, and your *Conclusion*. The blueprint for an *Abstract* is, "We did. We saw. We concluded," and your *Abstract* should include the background of or the reason for your study, the methods you used, a list of your main findings, and your conclusions.

An *Abstract* should be a single lean paragraph. It should be written in complete sentences. If it includes any technical abbreviations, then it should also include their definitions. *Abstracts* have no figures or tables, and *Abstracts* rarely cite references.

8.1.   The Simple *Abstract*

Scientific journals use two different forms for their *Abstracts*. Traditional journals use a simple *Abstract*—an undifferentiated paragraph of less than 200 words.

> For example, the simple *Abstract* for my axon staining paper was:
>
> • "**ABSTRACT**. Using a photographic enhancing solution, we were able to intensify the Bodian nerve stain. Intensifying Bodian-stained fixed cells and tissues improved the detail that can be resolved and revealed immature axons and small cell processes. With our post-staining intensification technique, the fine details of growth cones in tissue culture could be seen clearly using standard light microscopy."

The following are a variety of simple *Abstracts*:

• From a report on using high doses of cytosine arabinoside (ARA-C) to keep patients with acute myeloid leukemia (AML) in continuous complete remission (CCR) (Bohm et al., 2005):

"**Abstract**

"High dose intermittent ARA-C ($2 \times 3\,g/m^2$ i.v., days 1, 3, 5) = HiDAC was introduced as consolidation in AML by the CALGB-group in 1994. We treated 44 de novo AML patients in CR with up to four cycles of HiDAC (four cycles:

56.8%; three cycles: 22.7%; two cycles: 6.8%; one cycle: 13.7%). Median duration of aplasia (ANC< 0.5 × 10⁹/l) was 12 days. Neutropenic fever occurred in 38.6% of the patients during the first, 52.6% during the second, 45.7% during the third, and in 40% during the fourth cycle. Non-hematologic toxicity was tolerable. The median overall-and disease-free survivals were 19.3 and 11.3 months, respectively. The best outcome was seen in patients aged <40 years. These results confirm that HiDAC is a safe and effective consolidation in AML.
"**Key Words**: AML; Consolidation chemotherapy; High dose ARA-C; Toxicity"

- From a report on the unmitigated cerebral suppression of sound from the ipsilateral ear in split-brain patients (Milner et al., 1968):

"**Abstract.** Right-handed patients with surgical disconnection of the cerebral hemispheres cannot report verbal input to the left ear if different verbal stimuli have been channeled simultaneously to the right ear. With monaural stimulation, they show equal accuracy of report for the two ears. These findings highlight the dominance of the contralateral over the ipsilateral auditory projection system. Suppression of right-ear input is obtained in nonverbal tests. Dissociation between verbal and left-hand stereognostic responses indicates a right-left dichotomy for auditory experience in the disconnected hemispheres."

- From a report on the effect on animal diets of S-methylmethionine, a compound found only in plants (Augspurger et al., 2005):

"**ABSTRACT** Acid hydrolysis of dehulled soybean meal (SBM) and corn gluten meal (CGM) followed by chromatographic amino acid analysis (ninhydrin detection) revealed substantial quantities of S-methylmethionine (SMM) in both ingredients (1.65 g SMM/kg SBM; 0.5 g SMM/kg CGM). Young chicks were used to quantify the methionine-(Met) and choline-sparing bioactivity of crystalline L-SMM, relative to L-Met and choline chloride standards in 3 assays. A soy isolate basal diet was developed that could be made markedly deficient in Met, choline, or both. When singly deficient in choline or in both choline and Met, dietary SMM addition produced a significant (P<0.01) growth response. In Assay 2, dietary SMM did not affect (P>0.10) growth of chicks fed a Met-deficient, choline-adequate diet. A standard-curve growth assay revealed choline bioactivity values (wt:wt) of 14.2 ± 0.8 and 25.9 ± 5.1 g/100 g SMM based on weight gain and gain:food responses, respectively. A fourth assay, using standard-curve procedures, showed choline bioactivity values of 20.1 ± 1.1 and 22.9 ± 1.7 g/100 g SMM based on weight gain and gain:food responses, respectively. It is apparent that SMM in foods and feeds has methylation bioactivity, and this has implications for proper assessment of dietary Met and choline requirements as well as their bioavailability in foods and feeds.
"**KEY WORDS**: choline, methionine, S-methylmethionine, betaine, chick"

- From a report of improvements to a model of the release kinetics of solute from a polymeric matrix, with experimental tests of this improved model (Paul and McSpadden, 1976):

"The theory of diffusional release of a solute from a polymeric matrix where the initial loading of solute is less than or greater than the solubility limit has been reviewed and extended. The conceptual model for the saturated case proposed by Higuchi has been refined to remove the inaccuracies caused by the 'pseudosteady-state' assumption. A finite external mass transfer resistance has also been incorporated into the present analysis. For all solute loadings, the asymptotic release profile is seen to be a linear plot of total amount of solute released versus the square root of time, which has a finite intercept on the square root of time axis because of the external resistance. The dependence of the slope and intercept on solute loading and other system parameters can be predicted for all cases with the models presented. Experimental release rates of an organic dye from a silicone polymer into acetone were measured for a range of solute loadings in order to test the applicability of these equations. All system parameters except the external mass transfer coefficient were measured by independent experiments. The experimental release data were described very well by the computed results."

- From a report examining the ultrasound echogenicity of the substantia nigra in Parkinson's disease (Berg et al., 2001):

"**Abstract** Recently an increased echogenicity of the substantia nigra (SN) in patients with Parkinson's disease (PD) was demonstrated by transcranial ultrasound (TCS). In this study we set out to compare SN echogenicity with disease characteristics (time of onset, duration, toxin exposure) in a large patients sample. Patients' history and exposure to toxins were recorded from 112 PD patients who underwent a thorough neurological examination including assessment of disease stage according to Hoehn and Yahr and CURS (Columbia University Rating Scale). Personality was assessed according to the Freiburg Personality Inventory. In all patients, the area of SN echogenicity was encircled and measured by TCS. All except 9 patients had hyperechogenic SN areas exceeding the mean plus standard deviation values of an age matched control group ($0.19\,cm^2$). The age of disease onset was lower in patients who displayed an area of SN echogenicity above this value. The area of SN echogenicity was larger contralateral to the side with more severe symptoms. None of the other characteristics correlated with ultrasound findings. We conclude that SN hyperechogenicity is a typical finding in PD. The cause of hyperechogenicity is so far unknown. Investigation of the underlying reason might disclose a pathogenic factor in PD.
"**Keywords**: Parkinson's disease, Transcranial ultrasound, Substantia nigra hyperechogenicity, Vulnerability marker"

- From a report on the effects of three natural agonists on the contraction of smooth muscle in lung airways (Perez and Sanderson, 2005):

"**ABSTRACT** Increased resistance of airways or blood vessels within the lung is associated with asthma or pulmonary hypertension and results from contraction of smooth muscle cells (SMCs). To study the mechanisms regulating these contractions, we developed a mouse lung slice preparation containing bronchioles and arterioles and used phase-contrast and confocal microscopy to correlate the contractile responses with changes in $[Ca^{2+}]_i$ of the SMCs. The airways are the focus

of this study. The agonists, 5-hydroxytrypamine (5-HT) and acetylcholine (ACH) induced a concentration-dependent contraction of the airways. High concentrations of KCl induced twitching of the airway SMCs but had little effect on airway size. 5-HT and ACH induced asynchronous oscillations in $[Ca^{2+}]_i$ that propagated as Ca^{2+} waves within the airway SMCs. The frequency of the Ca^{2+} oscillations was dependent on the agonist concentration and correlated with the extent of sustained airway contraction. In the absence of extracellular Ca^{2+} or in the presence of Ni^{2+}, the frequency of the Ca^{2+} oscillations declined and the airway relaxed. By contrast, KCl induced low frequency Ca^{2+} oscillations that were associated with SMC twitching. Each KCl-induced Ca^{2+} oscillation consisted of a large Ca^{2+} wave that was preceded by multiple localized Ca^{2+} transients. KCl-induced responses were resistant to neurotransmitter blockers but were abolished by Ni^{2+} or nifedipine and the absence of extracellular Ca^{2+}. Caffeine abolished the contractile effects of 5-HT, ACH, and KCl. These results indicate that (a) 5-HT and ACH induce airway SMC contraction by initiating Ca^{2+} oscillations, (b) KCl induces Ca^{2+} transients and twitching by overloading and releasing Ca^{2+} from intracellular stores, (c) a sustained, Ni^{2+}-sensitive, influx of Ca^{2+} mediates the refilling of stores to maintain Ca^{2+} oscillations and, in turn, SMC contraction, and (d) the magnitude of sustained airway SMC contraction is regulated by the frequency of Ca^{2+} oscillations.
"**KEY WORDS**: laser scanning, confocal microscopy, asthma, airways, arterioles, mouse lung slices"

8.2.   The *Abstract* with Subsections

Other journals use *Abstracts* that are miniatures of the actual paper. These *Abstracts* are usually longer (200–350 words) and are written in subsections that parallel the outline of the article. Typically, the subsections are:

- *BACKGROUND* (CONTEXT, OBJECTIVE) = 1–2 sentences
- *METHODS* (METHODOLOGY) = 2–3 sentences
- *RESULTS* = <10 sentences
- *CONCLUSION* = 1 sentence

These subsections are brief summaries of the paper's

- *Introduction*
- *Materials and Methods*
- *Results*
- *Discussion and Conclusion*

For example, the Abstract with subsections for my axon staining paper might have been:
- "*ABSTRACT.* **BACKGROUND**: The Bodian silver stain, a generally robust and specific stain for neurons, does not cleanly stain fine cell processes

(continued)

(continued)

> such as embryonic axons in situ and growing neurites in tissue culture. **METHODS**: We used a modified photographic enhancement procedure as a post-staining intensifier for the standard Bodian technique. **RESULTS**: Intensifying Bodian-stained fixed tissues cleanly highlighted small embryonic tadpole axons in spinal cord sections. In embryonic chick cell cultures, details of thin neurites and their growth cones were visible. **CONCLUSION**: With post-staining intensification of the Bodian silver stain, the fine details of immature axons could be seen clearly using standard light microscopy."

The following are two examples of *Abstracts* with subsections.

- From a report on the use of proton magnetic resonance spectroscopy to assess the severity of damage in a shaken infant (Haseler et al., 1997):

"**ABSTRACT. Objective.** The purpose of this study was to use proton magnetic resonance spectroscopy (MRS) as a metabolic assay to describe biochemical changes during the evolution of neuronal injury in infants after shaken baby syndrome (SBS), that explain the disparity between apparent physical injury and the neurological deficit after SBS. **Methodology.** Three infants [6 months (A), 5 weeks (B), 7 months (C)] with SBS were examined repeatedly using localized quantitative proton MRS. Examinations were performed on days 7 and 13 (A), on days 1, 3, 5, and 12 (B), and on days 7 and 19 (C) post trauma. Long-term follow-up examinations were performed 5 months post trauma (A) and 4.6 months post trauma (B). Data were compared to control data from 52 neurologically normal infants presented in a previous study. **Results.** Spectra from parietal white matter obtained at approximately the same time after injury (5 to 7 days) showed markedly different patterns of abnormality. Infant A shows near normal levels of the neuronal marker N-acetyl aspartate, creatine, and phosphocreatine, although infant C shows absent N-acetyl aspartate, almost absent creatine and phosphocreatine, and a great excess of lactate/lipid and lipid. Analysis of the time course in infant B appears to connect these variations as markers of the severity of head injury suffered in the abuse, indicating a progression of biochemical abnormality. The principal cerebral metabolites detected by MRS that remain normal up to 24 hours fall precipitately to 40% of normal within 5 to 12 days, with lactate/lipid and lipid levels more than doubling concentration between days 5 and 12. **Conclusions.** A strong impression is gained of MRS as a prognostic marker because infant A recovered although infants B and C remained in a state consistent with compromised neurological capacity. Loss of integrity of the proton MRS spectrum appears to signal irreversible neurological damage and occurs at a time when clinical and neurological status gives no indication of long-term outcome. These results suggest the value of sequential MRS in the management of SBS. "**KEY WORDS**: shaken baby syndrome, traumatic brain injury, neuronal injury, magnetic resonance imaging, magnetic resonance spectroscopy."

- From a report examining the effect of matrix metalloproteinase inhibitors on healing after periodontal surgery (Gapski et al., 2004):

"**Background**: The adjunctive use of matrix metalloproteinase (MMP) inhibitors with scaling and root planing (SRP) promotes new attachment in patients with periodontal disease. This pilot study was designed to examine aspects of the biological response brought about by the MMP inhibitor low dose doxycycline (LDD) combined with access flap surgery (AFS) on the modulation of periodontal wound repair in patients with severe chronic periodontitis.

"**Methods**: Twenty-four subjects were enrolled into a 12-month randomized, placebo-controlled, double-masked trial to evaluate clinical, biochemical, and microbial measures of disease in response to 6 months therapy of either placebo capsules + AFS or LDD (20 mg b.i.d.) + AFS. Clinical measures including probing depth (PD), clinical attachment levels (CAL), and bleeding on probing (BOP) as well as gingival crevicular fluid bone marker assessment (ICTP) and microbial DNA analysis (levels and proportions of 40 bacterial species) were performed at baseline and 3, 6, 9, and 12 months.

"**Results**: Patients treated with LDD +AFS showed more potent reductions in PD in surgically treated sites of >6 mm($P < 0.05$, 12 months). Furthermore, LDD + AFS resulted in greater reductions in ICTP levels compared to placebo + AFS. Rebounds in ICTP levels were noted when the drug was withdrawn. No statistical differences between the groups in mean counts were found for any pathogen tested.

"**Conclusions**: This pilot study suggests that LDD in combination with AFS may improve the response of surgical therapy in reducing probing depth in severe chronic periodontal disease. LDD administration also tends to reduce local periodontal bone resorption during drug administration. The use of LDD did not appear to contribute to any significant shifts in the microbiota beyond that of surgery alone.

"**KEY WORDS**: bone resorption/prevention and control; doxycycline/therapeutic use; periodontal attachment; periodontal diseases/therapy; surgical flaps."

9.   *KEY WORDS* AND THE *LIST OF NONSTANDARD ABBREVIATIONS*

9.1.   Key Words

Some journals ask that you follow your *Abstract* with a list of 3–10 key words or phrases. These terms will be used to index your article under standard headings in large databases. Therefore, besides choosing key words that characterize the specific focus of your paper, include some terms that categorize your paper more generally. List your key words alphabetically on a separate line after the *Abstract*.

For example, a paper on the number of dorsal stripes on Guatemalan tarantulas might have these key words:

"**Key Words**: *Brachypelma*, Central American spiders, coloration of spiders, dorsal stripes, Guatemalan arthropods, tarantulas"

The section on *Abstracts*, above, has other examples of **Key Words**.

9.2.   **Alphabetical List of Nonstandard Abbreviations**

Abbreviations make your writing compact. Standard abbreviations, such as 'm' for 'meter' and 'g' for 'gram', are part of the common language of science, and you can use them freely. **Appendix C**, at the end of this book, lists many of the standard scientific abbreviations and gives rules for their use.

In addition, each branch of science has developed its own shorthand for long technical terms; for example, medicine uses 'CHF' for 'congestive heart failure', and neurosciences uses 'NGF' for 'nerve growth factor.' Well-known as they may be in their own fields, many of these abbreviations are not yet standard for all science.

Any nonstandard abbreviations that you use as shorthand must be defined in the main text of your paper. The convention is to put the abbreviation in parentheses when it first appears in the text of your paper, such as, "Of the patients with congestive heart failure (CHF), 10% had ..." Thereafter, you can use the abbreviation alone, such as, "Fourteen of the patients with CHF had lower extremity edema."

In addition to putting definitions in the main text, you can help your reader by including an alphabetical list of the nonstandard abbreviations used in your paper; for example:

> "*Abbreviations*
> CHF congestive heart failure
> FPG fasting plasma glucose level
> NGF nerve growth factor"

In your manuscript, put the list of your nonstandard abbreviations at the end of the *Abstract*, just below the **Key Words**.

10.   **TITLE**

> *Skeleton of the* Title
>
> Complete Summary of the Paper in Two Lines or Less

Each scientific paper needs a *Title* and an *Abstract*. Together, the two form a small scientific report of their own, and the combination is used as a stand-in for the complete article in condensed databases, such as *Biological Abstracts* and *Medlines*.

By itself, the *Title* is the ultimate précis of your paper, so fill it with clear and useful information. Write,

• "Photographic Intensifier Improves Bodian Staining of Tissue Sections and Cell Cultures"

not the mysterious,

- "What Can Intensification Add to Bodian Stains?"

or the empty,

- "A New Way to Stain Nerve Cells"

When composing a *Title*, build it with words that characterize your entire paper, because *Titles* are often used for indexing articles. Begin by listing 8–12 terms that capture the essence of your recipes, results, and conclusions, and include the key variables that are the focus of your experiment. Next, arrange the words of your list into a complete phrase. Finally, rework the draft of your title so that it meets these basic requirements:

- It should recapitulate your *Conclusion*.
- It should be succinct. Limit your title to two or fewer lines of text.
- It should include a verb, which should be in the present tense. Find an active verb, and aim for the grammatical structure *Subject–Active verb–Object*, i.e., "Viral Interleukin-6 Blocks Interferon Signaling".
- It should not be thickly worded. Use no more than three modifiers for any one noun. For example, write, "Large Red Guatemalan Tarantulas Have 8 Stripes" not "Large Red 8-striped Guatemalan Tarantulas are the Most Common Variety".

Here are active *Titles* from a wide variety of recent scientific papers. I have included many examples with the hope that you can find a model that will be appropriate for your particular paper.

- "Sugar Feeding Reduces Short-term Activity of a Parasitoid Wasp"

- "Propanil and Swep Inhibit 4-coumarate:CoA Ligase Activity *in vitro*"

- "The Anti-schistosomal Drug Praziquantel is an Adenosine Antagonist"

- "Subcutaneous Tri-Block Copolymer Produces Recovery from Spinal Cord Injury"

- "The Tyrosinase Enhancer is Activated by Sox10 and Mitf in Mouse Melanocytes"

- "Individual Differences in Optimism Predict the Recall of Personally Relevant Information"

- "Dead Spaces Hinder Diffusion and Contribute to Tortuosity of Brain Extracellular Space"

- "Composite Carriers Improve the Aerosolisation Efficiency of Drugs for Respiratory Delivery"

- "Age and Nutrient-Limitation Enhance Polyunsaturated Aldehyde Production in Marine Diatoms"

- "The Dietary S-Methylmethionine, a Component of Foods, Has Choline-Sparing Activity in Chickens"

- "Mice Genetically Deficient in Neuromedin U Receptor 2, but not Neuromedin U Receptor 1, Have Impaired Nociceptive Responses"

- "RXRα Regulates the Pregnancy-Specific Glycoprotein 5 Gene Transcription Through a Functional Retinoic Acid Responsive Element"

- "Piperonyl Butoxide Induces the Expression of Cytochrome P450 and Glutathione *S*-Transferase Genes in *Drosophila melanogaster*"

- "One-Month Therapy with Simvastatin Restores Endothelial Function in Hypercholesterolemic Children and Adolescents"

- "RBP-J (CSL) is Essential for Activation of the K14/vGPCR Promoter of Kaposi's Sarcoma-Associated Herpesvirus by the Lytic Switch Protein RTA"

11. FOOTNOTES

> *Skeleton of a* Footnote
>
> A Superscript Number Followed By One Brief Paragraph

For most journals (*Science* is the prominent exception), footnotes are a minor part of their scientific articles.

Footnotes are only for information that is not necessary for your presentation. If your paper is incomplete without a certain fact, reference, or comment, then that information belongs in the main text. Sometimes, however, there are small side-lights that you would like to record for readers who will be going off on a tangent from the subject of your paper. Put these nuggets in footnotes.

> To record an event that might otherwise be overlooked or forgotten, you can use a footnote. For example, in a paper on the number of petals on the flowers of a newly developed black-eyed Susan hybrid, the following information should be in a footnote.
>
> "¹Like most members of the black-eyed Susan family, our hybrid *Rudbeckia* produces flowers with yellow petals. In the keynote address at the 2002 World Congress of Black-eyed Susan Growers, Prof. Max Katz-Breit of the University of Ostrow showed pictures of the first known hybrid *Rudbeckia* flowers with blue petals."

Each footnote must be cited in the body of the text, and any footnotes should be numbered in the order in which they are cited. When you submit your manuscript to a journal, list all the footnotes on a separate page (titled "Footnotes"), and put this page after the main text of your paper.

- Here is an example of a tangential comment appropriately put into a footnote in a classic scientific paper that shows the weight of radioactive lead isotopes to be different from the weight of ordinary lead (Richards and Lembert, 1914):

"³Mr. Max E. Lembert, Dipl. Ing., a pupil of Dr. Fajans, was sent by him and the Technische Hochschule of Karlsruhe, with the support of Professor Bredig, to

Harvard University especially for this purpose. Sir William Ramsay, also, at about the same time, had urged on behalf of Dr. Soddy that the atomic weight of radioactive lead should be studied in the Wolcott Gibbs Memorial Laboratory. It is needless to say that the opportunity was welcomed; indeed, the matter would have been taken up here before, except for a fear of trespassing upon a field which might properly be considered as belonging to the proposers of the theory. A brief announcement of this work was made by Dr. Fajans at the meeting of the Bunsen Gesellschaft in Leipzig on May 21st, and a brief notice was published in "Science" on June 5, 1914.'

12.   ACKNOWLEDGEMENTS

> *Skeleton of the* Acknowledgements
>
> One Paragraph

When your paper is finished, add one last historical note, the *Acknowledgements*. This is a paragraph that usually comes after the *Conclusion* and before the *References*.

The *Acknowledgements* is an addendum to the *Materials and Methods*. In simple complete sentences, it lists those people and institutions who gave you advice, information, assistance, and materials. It should also list all the sources of your financial support.

The *Acknowledgements* should be brief, however, the details are almost entirely in your hands. Here are some examples to illustrate the range of *Acknowledgements* found in the scientific literature:

- "This study was supported, in part, by Grant RG-5920 from the National Institutes of Health."

- "I desire to express my thanks to Mr. William Kay for his invaluable assistance in counting scintillations."

- "This work was made possible by a grant from the Nuffield Foundation, for which we should like to express our gratitude. Our thanks are also due to Professor J.Z. Young for his criticism of the manuscript."

- "We thank D. Wang, Y.T. Liu, and A. Urisman for their technical assistance on the microarrays. B. Glaunsinger is supported by an American Cancer Society postdoctoral fellowship, and D. Ganem is a member of the Howard Hughes Medical Institute. The authors have no conflicting financial interests."

- "Grateful thanks are due to J.D. Archibald, D.A. Eberth, K. Howard, J.A. Lillegraven, G.J. Retallack, and A. Sweet, all of whom reviewed earlier drafts of this manuscript and contributed immensely to its improvement. However, the viewpoints expressed here are not necessarily those of the reviewers. This work was supported in part by the Instituto de Geologia, Universidad Nacional Autonoma de Mexico, and by a Fulbright Garcia-Robles Award."

- "This study was supported by the Fonds zur Forderung der Wissenschaftlichen Forschung in Osterreich (FWF) grant #P-14031. Contributions. Alexandra Bohm contributed to collection, analysis and interpretation of data and drafting of the article, Maria Piribauer contributed to data collection and assembly, Friedrich Wimazal contributed to collection of data, Klaus Geissler contributed patients and logistic support, Heinz Gisslinger contributed patients as well as administrative support, Paul Knobl contributed patients and administrative support, Ulrich Jager contributed patients and administrative as well as logistical support, Christa Fonatsch contributed analysis and interpretation of cytogenetics, Paul A. Kyrle contributed patients and logistical support, Peter Valent contributed to the conception and design of the study, interpretation of data, drafting, and final approval, Klaus Lechner contributed administrative and logistical support and critical revision of the article, Wolfgang R. Sperr contributed conception and design of the study, statistical analysis and interpretation of data as well as drafting and final approval of the article."

13.   REFERENCES

> *Skeleton of the* References
>
> A List of All Sources Cited in the Paper,
> using the Appropriate Bibliographic Format

Scientific observations (i.e., *recipe* → *results* reports) are linked together in a complex network of scientific articles. The *References* section uses a standardized bibliographic format to list significant links through which your paper—your *recipe* → *results* report—is tied into the web-like archives of science.

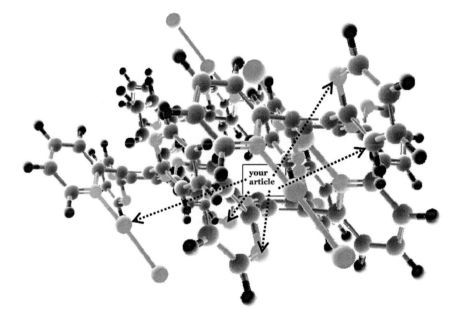

In a published paper, the *References* section comes after the *Conclusion* and the *Acknowledgements*. A thorough reference list should include citations to:

- Background information
- The sources of your experimental techniques
- Reports of similar observations
- Discussions of proposals or generalities related to those you have made in your *Discussion*
- Papers with similar *Conclusions*

13.1.   Cite Original Sources

Compiling your *References* is an opportunity to be sure you have dug into the relevant literature. Therefore, when you are piecing together your list of citations, do some extra library work using the extensive bibliographic search power of the Internet. As you hunt for connected scientific papers, remember to look for contrasting as well as supportive reports.

At the same time, make your final reference list useful by being selective. Cite only sources that relate directly to your particular region of study, so that your bibliographic list is deep but not overly broad.

- For links to background information, you do not have to cite the entire history of the subject. Instead, narrow your *References* to those books and papers that will give readers an entryway to the key literature.

- For links to information about directly related research studies, cite the original papers. Avoid citing review articles and general books. Summaries and reviews do not have the space to include all the original data with its full range of quirks, qualifiers, and special cases, and summary articles frequently miss important information – occasionally they even get the facts wrong. The best use of broad reference articles is as a way for you to identify the major original scientific reports, which you should then read and cite.

13.2.   Don't Cite Incomplete Sources

Don't cite abstracts, because these often contain preliminary information and early speculation based on incomplete data. Also, check that papers "in press" have already been accepted and are scheduled for publication.

Avoid citing "personal communications." However, if you feel that this form of information is essential to your report, list the full name and address of the source and the date you received the information and provide the journal with written permission from the source.

13.3. Find the Appropriate Citation Formats

In any scientific paper, you will use two formats for your citations:

- The complete bibliographic format is used in the *References* section at the end of your paper.
- A brief citation format is used in the body of the text.

13.3.1. Bibliographic Format

In the *References* section, citations will have formats similar to these:

- Gorkovic CM. *Transplant recipients with AIDS enlarge T cell populations post-operatively.* [dissertation]. Boston: Willoughby Univ; 2007.
- Halpern SD, Ubel PA, Caplan AL. Solid-organ transplantation in HIV-infected patients. *N Engl J Med.* 2002; 347(4): 284–7.

13.3.2. Citation Format

In your text, citations often have the format '(Halpern et al., 2002)', as in

- "… A recent study of transplantation in immune compromised patients (Halpern et al., 2002) found that 65% of kidney recipients …"

However, some journals number all the bibliographic entries in the *References* section, and the text citations are simply the entry numbers. For example, the *References* section might look like:

- 1. Gorkovic CM. *Transplant recipients with AIDS enlarge T cell populations postoperatively.* [dissertation]. Boston: Willoughby Univ; 2007.
- 2. Halpern SD, Ubel PA, Caplan AL. Solid-organ transplantation in HIV-infected patients. *N Engl J Med.* 2002; 347(4): 284–7.

and the text citation would look like:

- "… A recent study of transplantation in immune compromised patients (2) found that 65% of kidney recipients …"

13.3.3. Journal Formats Vary

The exact styles of the bibliographic and text citation formats vary from journal to journal, so consult the *Information for Contributors* section of your target journal before finalizing your *References* section. Many medicine-related journals use the same formats, as detailed in the *Uniform Requirements for Manuscripts Submitted to Biomedical Journals* (ICMJE requirements), which are available online at *www. icmje.org.* In the special cases of theses and dissertations, follow the formats in Turabian, KL et al. 2007. *A Manual for Writers of Research Papers, Theses, and Dissertations,* 7th ed. University of Chicago Press, Chicago, IL. Additional style

manuals are listed in **APPENDIX E** below, and samples of the ICMJE bibliographic formats are listed in **APPENDIX D**.

13.4.   Use Bibliographic Software

During your research, you should record and annotate all your reference
sources in your computerized notebooks. Bibliographic software packages are ideal
tools for building a library of sources. Good archiving programs include *EndNote*,
ProCite, *Reference Manager*, and *Zotero*. Later, when you have chosen the specific
journal to which you will be submitting your manuscript, the software can easily
adjust your reference list to the appropriate bibliographic format.

13.5.   Check Your Text Citations and Your Reference List

As you rearrange and rewrite your manuscript, the text citations will get pushed
around, changed, and dropped. Therefore, after you finish polishing your manuscript, take a few minutes to check the text citations. Make sure that each of the
bibliographic entries in the *References* is still cited somewhere in the text or the
figure legends.

To make this check, use your bibliographic software to print out a full copy
of your *References* section in an easy-to-read format. Then, take the copy and go
through the manuscript from beginning to end. When a bibliographic entry is cited,
mark it on your copy of the *References*. In the end, add any citations that were
inadvertently dropped from the *References* and delete any bibliographic entries no
longer cited in your text.

The *Uniform Requirements for Manuscripts Submitted to Biomedical Journals:
Writing and Editing for Biomedical Publication*—updated February 2006 (found
at: *http://www.icmje.org/*) adds this important reminder:

• "Some journals check the accuracy of all reference citations, but not all journals
  do so, and citation errors sometimes appear in the published version of articles.
  To minimize such errors, authors should therefore verify references against the
  original documents. Authors are responsible for checking that none of the references cite retracted articles except in the context of referring to the retraction."

Part III
PREPARING A MANUSCRIPT FOR SUBMISSION

Chapter 1

CHOOSING A JOURNAL

Writing is tiring. When you have exhausted yourself on a draft, put the whole project aside, and relax for a day, a week, or a month.

Let the details fade and the immediacy subside. Think about other things until the studies, the experiments, and the manuscript sink into that subconscious womb where they can gestate peacefully and can begin to assume a balanced form.

1.   MAKE A LIST OF CANDIDATE JOURNALS

Sometime during your writing or while your unfinished manuscript gestates, choose your target journal. The general format for papers is standard throughout science, but the specific skeletons differ from field to field. Even within a field, each scientific journal can have its own organizational style. Select a target journal before you have finished polishing your paper. This way, your final manuscript will have the appropriate organization and style.

Your list should have 4–6 candidate journals. Through your reading of the scientific literature, you will already have identified a number of journals that publish articles similar to yours. Pick three of these journals. Then, ask colleagues to suggest appropriate journals—if these journals are not already on your list, add them.

M.J. Katz, *From Research to Manuscript,*
© Springer Science+Business Media B.V. 2009

1.1.   Consider Open Access Publishing

In the past, only printed media had the permanence to be archived, and the public records of science were physically housed in libraries. Today, however, digital electronic media are being widely archived, and through projects like the National Library of Medicine's *PubMed Central*, digital articles are freely accessible via the Internet.

There are already a great many Internet-only scientific journals, and anyone with access to the worldwide web can read the articles in most of these journals. The cost of maintaining Internet journals is paid, in part, by a fee charged to the authors. (Many print journals also charge a publication fee to authors.) This kind of open access publishing makes scientific information available as a shared human heritage rather than as a commodity restricted by economics or politics.

In general, articles published in open access journals are read, used, and cited more often than those published in restricted access and printed journals. The main reason that authors hesitate to publish in Internet-only journals is that these journals have not yet achieved the prestige of some of the traditional scientific journals. Nonetheless, quality Internet-only journals adhere to rigorous acceptance standards and publish only peer-reviewed research articles.

Besides being widely accessible, good Internet-only journals publish their articles quickly, without the 3–6 month delay that paper journals require to print and to distribute each issue. Most Internet-only journals also let the authors retain the copyrights to their articles—another feature uncommon in print journals.

Information about open access biomedical journals can be found at:

• *http://www.pubmedcentral.nih.gov/about/intro.html*

A directory of open access journals can be found at:

• *http://www.doaj.org/*

Two large open access publishers are:

• Public Library of Science (PLoS) *http://www.plos.org/*
• BioMed Central *http://www.biomedcentral.com/*

Search for 1–3 open access journals appropriate for your paper, and add them to your list of candidates.

1.2.   Rank the Journals by Quality

Now rank your candidate journals according to quality. 'Quality' is, of course, subjective, but you can make a reasonable ranking by combining three assessments.

(a) *Ask colleagues to rank the journals on your list.*

(b) *Compare the acceptance rates of the journals.*

A lower acceptance rate (i.e., a higher rate of rejection) indicates a more critical review process. Some journals publish acceptance rates on their websites, but there are no central collecting sites for these statistics. The Ohio State University Extension's *Reference Guide of Professional Journals/Publications* lists the acceptance rates of selected journals at:

* *http://www.ag.ohio-state.edu/~admin/publication.htm*

In addition, a few other specific collections of acceptance rates are available online.

* For communication studies:
  o *http://www.natcom.org/nca/Template2.asp?bid=222*
* For education studies:
  o *http://cimc.education.wisc.edu/ed_info/journal%20acceptance%20rates.htm*
* For journals of the American Psychological Association:
  o *http://www.apa.org/journals/statistics.html*

(c) *Compare the* Impact Factor *of the journals.*

* The *Impact Factor* is a number used to estimate how often the articles of a journal are cited. The *Impact Factor* of a journal is the average number of times each of its articles is cited in the 2 years after publication. An *Impact Factor* of 1 means that, on average, each article in a journal was cited once within the 2-year period after its publication. Higher *Impact Factors* suggest that the journals are widely read, because their articles are cited frequently.

A higher *Impact Factor* often indicates a more critical or more prestigious journal. The full list of *Impact Factors* is found at the *ISI Web of Knowledge* (Thompson Corp.), which requires a subscription but which is available at most major libraries.

1.3.   Aim for the Best and Toughest Journal

It is natural to tire of the long, drawn-out process of writing a paper. By now, you have finished the lab or field work for your project, and you would like to publish your results and get on to new challenges. You are, however, facing the inevitable delay of waiting for reviews from a journal. A high-quality journal is more likely to ask for major revisions, and if you send your paper to the best and the toughest journal, even longer delays are inevitable. It can be tempting to cut the delays by submitting your manuscript to the least demanding journal.

The reason to accept the possibility of more delays until publication is that a critical review will improve your paper. Because you are the researcher, you will find it difficult to write a paper that will be clear to a broad audience. You know the subject of your research better than most of your readers. Your experiments have been a part of your daily life. You believe in your data, and you understand what your observations mean. Moreover, this is all history—you are already planning your next project.

Your future readers have not yet caught up with you. They are still living in the past, in the world as it was before your research. To write a paper that is clear to your audience, you must go back and recreate your pre-research mindset. You have to recall which of the things that currently seem obvious to you still need to be explained to an outsider.

Here is where reviewers can be indispensable. Reviewers are your ideal readers. They take the time to read the entire manuscript, and they put in extra effort to make sense of your writing. If a thoughtful reviewer does not understand your reasoning or has questions about your data, then your future audience will probably have the same difficulties and your manuscript (or your experiment) needs improvement. The best help will come from reviewers at the toughest journals, because they will not accept weak or unclear papers.

When you first send out your manuscript, it will not be perfect. Improve it by submitting it to the best and toughest journal on your list. You learn nothing from reviews that say, "This paper can be published as is." Instead, the most helpful review says, "Here are some specific things that will make this paper better."

2.   STYLE RULES

After choosing a journal, get a copy of its article format, i.e., its style rules. These rules have headings such as:

- *Requirements for Manuscripts*
- *Instructions for Contributors*
- *Information for Authors*
- *Notice to Contributors*
- *Author Guidelines*

To help biomedical authors, the International Committee of Medical Journal Editors (ICMJE) has formulated a set of widely accepted standards for articles. These rules are used by more than 600 journals. The ICMJE's *Uniform Requirements for Manuscripts Submitted to Biomedical Journals* can be found online at *www.icmje.org*.

Another general source for Style Rules is the *Instructions to Authors in the Health Sciences* website at *http://mulford.meduohio.edu/instr/* maintained by the

University of Toledo, Health Science Campus. This website has links to the style rules of more than 3,500 journals that publish in the health and life sciences.

When you find the manuscript guidelines for your target journal, put a copy on your desk, alongside a general style manual (see **Appendix E**, at the end of this book, for suggestions), and a dictionary.

Chapter 2

A FINAL REWRITE

After a rest, and when you feel fresh, bring out your manuscript again. Before you face the content, adjust the form to fit the style requirements of your target journal. Here are some specific things you will want to do:

- Rewrite the *Abstract* in the form used by your journal.
- Choose keywords to append to the *Title*, if these are required.
- Match figure legends to the journal's style.
- Change text citations to the appropriate style.
- Translate your *References* into the correct bibliographic format.

Now, attack the substance of the paper anew.[3] Go through the steps outlined in the chapters above, improve the figures, check the tables, make the order of the sentences more logical, and shorten the *Abstract* and the *Introduction*. Ask yourself, "Where do I read things over twice?" and "Where am I uncomfortable?" You will naturally feel those spots that are poorly written, weakly argued, confusingly phrased, and meagerly researched.

Inevitably, you will resist facing these difficult areas of the manuscript. You resist because, through your many reworkings, you have already found these places hard to fix. Steel yourself, and improve the problem areas, one by one.

When you tire of the details, reexamine the structure of the paper. Begin by reading through the headings. Could you use more headings? Are long subsections broken into enough separate parts? Are the headings symmetrical, i.e., if you have a "Measurements Before Treatment," do you also have a "Measurements After Treatment"?

---

[3] Will the rewriting never end? When I was a student and had not yet attempted any science writing, Kenneth Warren, a professor of mine, told me this harrowing story: 'I already had my MD and was doing a fellowship at the National Institutes of Health, studying liver disease. I brought the draft of my first manuscript to my boss. The next day, he handed it back to me with big question marks and X-ed out paragraphs. "Re-write this," he said. Over and over, I gave him clean new drafts and he gave me back worn, thickly penciled pages. More than 20 drafts came and went this way. Even after we sent the paper off for review, I found penciled comments on my desk. I vowed never to treat my students or post-docs so critically. And now? Twenty rewrites is what it takes for every one of my students, post-docs, and even research fellows. I've found it's inescapable.' At the time, I thought that Dr. Warren was exaggerating. I can now assure you, he was not.

M.J. Katz, *From Research to Manuscript,*
© Springer Science + Business Media B.V. 2009

Each time you return to your manuscript, be satisfied with making a few improvements, and then set the draft aside.

1.   GET A FRIENDLY CRITIQUE

When you feel that you have done all that you can, it is time for an independent perspective. Find a friendly reader, and ask for a critique. Besides making general comments, ask your reader to mark the places where he or she stumbles or stops to reread a sentence—these are sure to be problem areas. Afterward, listen to the report without interrupting to explain or to justify. Then go fix the manuscript.

2.   READ THE PAPER BACKWARDS

To learn to draw pictures of the real world, students are often advised to look at a scene upside down, through a grid, or otherwise broken into very small pieces.

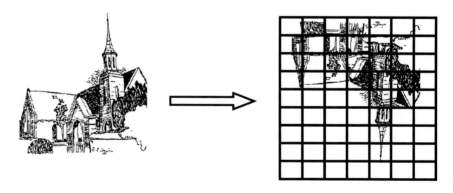

Normally, we do not consciously attend to many of the actual details of edge, contour, and shadow. Instead, we create a vision of objects and scenes as wholes. Our brains give a scene meaning by adding depth, revising edges, and tempering variations in lighting. Much of what we think we see is a creation of our mind.

When bits of a scene are taken out of context, however, they become unfamiliar and they revert to abstract patterns of light, dark, line, and shape. In an abstraction, we can see the details separately and starkly. They register simply as themselves, and they become easier to copy.

By this time in your writing, your manuscript has become too familiar. Your brain has built a picture of your paper, but that picture may not yet be clearly written on the page. When you read the manuscript, you jump past the details of the wording and instead you read sentences and paragraphs as wholes. You are desensitized to awkward and unclear phrases and to missing steps in logic. You already know what you mean to say, and you see the meaning even when it is imperfectly written.

It is time to make your manuscript unfamiliar. Turn it upside down, by reading it backwards, sentence by sentence.

Read Your Manuscript from the End to the Beginning

A. INTRODUCTION
    Scientific articles are the repositories of scientific observations, and they give the recipe by which those observations can be repeated.

    Scientific papers have a stereotyped format:
Abstract, Introduction, Materials and Methods, Results, Discussion, Conclusions, References.

B. MATERIALS AND METHODS
    Within this stereotyped format, the language of a scientific paper should be clean, clear, and unemotional. Much of the color of our everyday language derives from ill-defined, emotionally charged, ear-tickling images.

C. RESULTS
    Beyond a stereotyped format and clarity of language, a scientific paper also needs clarity of direction.

D. DISCUSSION
    A scientific paper must be useful to others. Having your paper reviewed by experts ensures that it can be understood and used by a broad scientific community.

E. CONCLUSION
    The scientific community will have the chance to use it.

Take each sentence on its own, out of context, and examine it alone. Make sure the sentence is clean and to the point. Replace long words by short ones and vague words by specific ones. If the sentence is too long, cut out words or break the sentence in two. When reading your paper backwards, your logic will feel strange, and this will force you to check whether the sentence that you have just read follows logically from the sentence that precedes it.

Work your way through every section, and make your paper read crisply upside down, as well as right-side up.

3.   RECHECK THE SPELLING

One of the great mysteries of writing is how, after ten rereadings, my manuscripts still contain misspelled words. Misspellings and typographic errors make a paper look like it was written carelessly and sent out hastily. Use a computer spell-checker before you send your manuscript to a journal.

Chapter 3

PREPARING AND SUBMITTING THE MANUSCRIPT

1.   PRINT AND PAGE FORMAT

When you send your finished manuscript to the journal, it should be in a form that can be directly used by the editors. The following guidelines apply to both physical and electronic manuscripts.

- *Font.* Use a clean, standard font. Appropriate, easy-to-read fonts include Arial, Microsoft Sans Serif, Tahoma, Times Roman, and Verdana.
- *Line Spacing.* Double space all the text, including the *References*, *Footnotes*, and *Figure Legends*. Also double space manuscripts submitted as electronic files.
- *Margins.* Leave large margins. For a standard 8.5"×11" page, the top, bottom, and both side margins should be 1–1.25" wide.
- *Page Numbers.* Number every page in your manuscript packet. Add an identifier to the number, such as "Katz, page 18."

2.   THE MANUSCRIPT PACKET

2.1.   Page One

The first page of your manuscript packet is the identification page. It should include the article title, the full names of the authors and their institutional affiliations, and the name and the contact information of the author taking responsibility for interacting with the editors.

The identification page should look like this:

M.J. Katz, *From Research to Manuscript,*
© Springer Science + Business Media B.V. 2009

Photographic Intensifier Improves Bodian Staining of Tissue Sections and Cell Cultures

Michael J. Katz and Lynne F. Watson
Department of Anatomy
Case Western Reserve University
Cleveland OH USA

correspondence:
Dr. Michael Katz
Dept. of Anatomy
Case Western Reserve University
10900 Euclid Ave.
Cleveland OH 44106-4977
USA
mjk8@case.edu
tel. 216-368-2390

Katz, page 1

2.2.   Page Two

The second page has a summary of the manuscript packet and any necessary disclosures, qualifications, and disclaimers.

2.2.1.   Summary

The summary includes the number of pages in the manuscript packet, the number of words in the main text, the number of figures (itemized as tables, graphs, drawings, and photographs), and the number of additional items (footnotes, appendixes, etc.), if there are any. For example:

> "**Manuscript:**
>   pages = 12
>   words = 1451
>   tables = 0
>   drawings = 1 (black-and-white)
>   photographs = 3 (half-tone)"

2.2.2.   Disclosures

The disclosures, qualifications, and disclaimers begin with the sources of funding for the research project. Next, list any non-authors who helped write the manuscript and give the source of the funds that were paid to these people. In addition, many journals now require a *Conflict of Interest* statement, which belongs in this section—conflicts of interest are any affiliations or funding sources that have the potential to bias the authors.

For example:

> "**Funding:**
> "A grant to MJ Katz from the Whitehall Foundation.
>
> "**Conflicts of Interest:**
> "The authors have no conflicts of interest with regard to this research or its funding."

2.3.   Page Three

The third page is the *Abstract*, with **KEY WORDS**, if this is the journal's style.

2.4.   Page Four

The manuscript proper begins on page 4 of the manuscript packet.

2.5.   The Last Pages

- At the end of the main text of your paper and after the *References* section put any *Appendices*.
- Next, put a page for any *Footnotes*.
- Following the *Footnotes*, put each Table and Figure on a separate page.
- End the packet with a page for all the Figure Legends.
- Number every sheet included in the manuscript packet.

Finally, when you are sending a paper version (as opposed to electronic files) of your manuscript, be sure to include as many copies as are required by the journal.

3.   THE INTRODUCTORY LETTER

Send your manuscript packet with a short, formal cover letter. In the letter, give:

- The full title of your article.
- The type of article (original research paper, brief report, review article, commentary, or letter).
- The essential point of the article (usually, the main conclusion).
- All the authors' names.

- A statement that each author has read the article and takes responsibility for its content.
- Details of any possible conflicts of interest.

For example

Dr. Janet Jones, Editor
Stain Technology

May 24, 2008

Dear Dr. Jones,

Please consider the accompanying original research manuscript "Photographic Intensifier Improves Bodian Staining of Tissue Sections and Cell Cultures" by Michael J. Katz and Lynne F. Watson for publication in *Stain Technology*. In this paper, we describe a chemical intensification procedure for a classic silver-based nerve stain. Both authors have read and approved the manuscript and take full responsibility for its content. The authors have no conflicts of interest in regard to this research or its funding.

Sincerely,

Michael J. Katz

Dept. of Anatomy
Case Western Reserve University
10900 Euclid Ave.
Cleveland OH 44106-4977 USA
mjk8@case.edu
tel. 216-368-2390

With your introductory letter, include copies of any permissions that you obtained to reproduce material from other sources.

Chapter 4

RESPONDING TO EDITORS AND REFEREES

You have sent a polished version of your manuscript to a journal, and after a few months, you will receive a packet with an editor's letter and anonymous reviews by 2–4 referees. Each review will be a critique that includes an overall evaluation and a list of items that need improving. Based on the reviews, the editor's letter will put your paper into one of three categories:

- The manuscript is accepted, pending specified changes.
- The manuscript requires revision and re-review.
- The manuscript is rejected.

You should have started the road to publication by submitting your manuscript to a demanding journal, one with a high rejection rate. If your manuscript is rejected, use the referees' comments to make improvements, and submit the paper to another journal.

1. HOMEWORK

After all your work, stay confident. You are going to publish your paper in this journal or in some other. Regardless of the editor's letter, use this as an opportunity to make improvements. Even good papers can use more work.

Therefore, before you face the details of the referees' comments, assign yourself some homework, beginning with reading. Find a new review article on the subject of your research, a review that you have not yet studied. Read the article and jot down each point that differs from what you have written, that agrees with what you have written, that expands on what you have written, or that fills a gap in what you have written. For every one of these points, make a change in your manuscript:

- *Differences.* If the article says something different from your manuscript, either change your manuscript or add a note such as "Smith et al. (2008) present an alternative view."
- *Agreements.* If the article cites a new supportive reference, consider adding it to your manuscript.
- *Further detail.* If the article includes relevant information beyond what you have written, add a summary statement to your manuscript with full references.
- *Gap-fillers.* If the article presents relevant information that you did not know, add the information to your manuscript, with references.

2.   THE COMMENT-BY-COMMENT LETTER

Now read the reviews carefully.

Begin making the suggested changes by drafting the letter that will accompany the manuscript back to the editor.

Even if your paper has been rejected, you can resubmit it to the same journal with a thoughtful letter. The review process should improve manuscripts, and a rejection is nothing to be embarrassed about. Show the editor that you have taken the criticisms seriously and have learned from them but that you still have confidence in your basic research.

The return letter should describe the changes you have made. In the first paragraph, summarize your initial repairs—the modifications that you made after reading the new review article—although those changes may have originated from you and not the reviewers.

> For example:
>
> August 18, 2004
> "Dear Editor,
>
> "I am enclosing a revised copy of my manuscript "Guatemalan Tarantulas Have 6–9 Dorsal Stripes."
>
> "I followed the reviewers' 8 suggestions. I have also taken this opportunity to improve the writing in the Discussion and to change the order of the tables in the Results. The Introduction cites 2 new supporting references. All told, the revised version is 3 paragraphs shorter and includes 4 new references. ..."

The remainder of the letter should itemize the specific improvements made in response to the reviewers' comments. As you write the letter, take the reviews comment by comment. If the suggestion makes sense to you, make the indicated change in the manuscript.

After making a change, write a sentence in your letter repeating the referee's suggestion and describing your response. Number each change.

For example:

"Here are the details of my responses to the reviewers' suggestions:
(1) Reviewer A said it was unlikely that I could infer an effect from only 16 cases. Therefore, I have added the results of a t-test, showing that the likelihood that my data are chance variants of a normal distribution is less than 5%.
(2) Reviewer A also pointed out that the second paragraph in the Discussion was wordy and imprecise. I have tightened the logic, taken a clearer stand, and in the process removed more than 20 words.
(3) Reviewer B said that tarantulas with 9 stripes have never been seen. Rawski (1943) reported seeing both 8-and 9-striped spiders on his wartime expedition to Guatemala. I have added this reference to my Introduction. ..."

When the letter is finished and the manuscript has been revised, let them both sit for a few days and then reread them before sending the packet back to the editor.

If you are sending the rewritten manuscript to a new journal, send essentially the same letter and include the previous editor's and reviewers' comments along with your responses. Good editors will respect an honest and open presentation.

3.   STAY CALM

Be polite in your letter. Most reviewers write with the best of intentions, but they rarely have the time to produce thorough, well thought out, error-free critiques. Some of the suggestions will not be useful. Pass over these comments. Also, ignore any emotion injected by the reviewer. All scientists have been hit with criticisms that sting. A referee once wrote of one of my manuscripts: "If this paper is published, it will set science back 50 years." The paper was eventually published in another journal, and science has survived. The best thing to do with a comment such as this is to use it as an anecdote for the end of your book on science writing.

Appendix A

WORDS THAT ARE OFTEN MISUSED

Ad hoc *adjective*

'Ad hoc' means 'for this special purpose.' An ad hoc committee is a committee specially convened for a particular purpose. Ad hoc assumptions are assumptions chosen expressly for the situation at hand. 'Ad hoc' does not mean 'temporary,' 'casual,' or 'without substantial basis.'

Affect/Effect *verb or noun*

'Affect' is usually a verb, as in: "Sunlight affected her mood."

'Effect' is usually a noun, as in: "Her cheerful mood was the effect of sunlight."

Alternate/Alternative *adjective or noun*

'Alternate' means 'every other member of a series' as in: "We assigned alternate hospital admission patients to the control and the treatment groups."

'Alternative' means 'another mutually exclusive possibility' as in: "We chose an alternative drug protocol." Don't use the vague expression 'viable alternative.' Be precise. Replace 'viable' with a more informative adjective, as in: "The only available [injectable, non-toxic, legal, already prepared, water-soluble, nonmagnetic, indelible, luminescent, inexpensive, non-zero, ...] alternative was ..."

All together/Altogether *adverb*

'All together' means 'all at once or all in one place' as in: "The four toddlers ate lunch all together [or just 'together']."

'Altogether' means 'when considered as one group or when taken together' as in: "Altogether, the four toddlers ate six cookies."

Arbitrary *adjective*

'Arbitrary' means 'without plan or design.' If you choose your experimental subjects by blindly pulling them from a pool of candidates, you are choosing them arbitrarily, not randomly. Truly random choices require very specific rules. (See **Random** below.)

Classic/Classical *adjective*

'Classic' means 'outstanding example' as in: "Albert Lehninger wrote a classic biochemistry textbook."

'Classical' means 'pertaining to Greek and Roman culture' as in: "In section one, Lehninger uses contemporary data and classical reasoning to explain the origins of alchemy."

Comprise/Compose/Constitute/Include *verb*

'Comprise' is transitive and its subject is a container. 'Comprise' means 'embrace' as in: "My circle of friends comprises only a guinea pig and a dog." [Write 'comprises' not 'is comprised of.'] When you use 'comprise,' you are talking about all the elements of the container.

'Include' is transitive and its subject is also a container. 'Include' means the more general term 'contain' as in: "My circle of friends includes a guinea pig." When you use 'include,' you can talk about only some elements of the container.

'Compose' and 'constitute' are transitive and their subjects are the containees, that is, the elements in the container. For example, "A guinea pig and a dog compose [or 'constitute'] my circle of friends."

Conjecture/Guess/Presume/Speculate/Surmise *verb*

'Conjecture,' 'speculate,' and 'surmise' are similar. These terms all mean 'offer a conclusion from incomplete evidence' as in: "After a brief visit to the crime scene, Holmes conjectured ['speculated' 'surmised'] that the killer was a neighbor."

'Presume' means 'conclude from previous knowledge or experience' as in: "From those footprints, Watson, we can presume that the killer was less than five feet tall," said Holmes.

'Guess' means 'offer a haphazard proposal' as in:

"Holmes," said the doctor, "what would you say if I suggested an extra-terrestrial hand had been at work here?" "I would say, my dear Watson, that you have made a rather poor guess."

Constantly/Continually/Continuously *adverb*

'Constant' means 'steady and unceasing' as in: "The addition of NaCl produced luminescence constantly for 2 min."

'Continual' describes a fragmented stream of discrete events and means 'repeating over and over again' as in: "The addition of KCl produced irregular bursts of luminescence continually for 1 min."

'Continuous' describes an unbroken stream of discrete events and means 'uninterrupted and without pause' as in: "The addition of LiCl produced a steady series of 0.5 sec flashes of luminescence continuously for 5 min."

Correlate *verb*

'Correlate' means 'relate together.' As a verb, this term tells us very little. Avoid empty sentences about relationships using the word 'correlate' as in, "The cellular effects of drug A correlated with those of drug B." Instead, specify the particular co-relationship that applies. Write: "The cellular changes caused by drug A were identical to those caused by drug B" or "The sequence of cellular effects of drug A and drug B were the same" or "The cellular effects of drug A and drug B occurred simultaneously" or "Drug A and drug B each interacted with the same membrane molecules" or "Drug A and drug B each produced the same cellular toxicity."

Data *noun*

'Data' has become both a singular and a plural noun—no longer is there a 'datum.'

Deduce/Induce/Infer *verb*

'Deduce' means 'particularize' or 'predict a specific example' as in: "From principles in Darwin's The Origin of Species, we can deduce that frogs and salamanders had common ancestors."

'Induce' means 'generalize' or 'predict a larger principle' as in: "Darwin used a variety of observations including the domestic breeding of dogs to induce the principles in The Origin of Species."

'Infer' is all encompassing. It means only 'conclude through logical reasoning' as in: "Somehow, Darwin inferred that the vertebrate eye evolved from simpler precursors, although those precursors were unknown in his day."

Endogenous/Inherent/Innate/Intrinsic *adjective*

'Endogenous' has an implied action. It means 'produced or originating from within' as in: "Euphoria can result from endogenous opioid-like hormones produced in the brain."

'Inherent,' 'innate,' and 'intrinsic' all refer to properties built into the system:

'Inherent' identifies a permanent property, as in 'inherent elasticity of rubber.'

'Innate' is used for living things, as in: 'innate logic of the mind' (but 'inherent logic of a computer chip').

'Intrinsic' usually describes a variable property, such as 'intrinsic body temperature' or 'intrinsic metabolic rate.'

Enhance/Increase *verb*

'Enhance' means 'heighten the contrast' as in: "We enhanced the resolution of the image" or "Well-chosen chords enhanced the clarity of the melody."

'Increase' means 'enlarge the value' as in: "We increased the brightness of the image" or "The addition of trumpets increased the volume of the melody."

Incidence/Prevalence *noun*

'Incidence' is the number of new cases, as in: "The incidence of type I diabetes in the U.S. is 30,000 cases each year."

'Prevalence' is the total number of cases at the moment, as in: "In 2003, the prevalence of type I diabetes in the U.S. was 1.3 million cases."

Lay/Lie *verb*

'Lay' (lay, laid, laid, laying) is transitive. It means 'place something' as in: "We laid the tubing along the ridge."

'Lie' (lie, lay, lain, lying) is intransitive. It means 'recline:' "The tubing lay along the ridge."

Necessary/Sufficient *adjective*

'Necessary' means 'absolutely required for a certain result' as in: "Drug A is necessary—without it, patients will not recover."

'Sufficient' means 'by itself will produce a certain result' as in: "Drugs B and C are each a sufficient therapy—either drug will cure the patient."

None *pronoun*

'None' can mean 'not any' or 'not one.' 'None' can take either a plural or a singular verb, depending on the context. In other words, you can write, "None of the electrodes were bent by the insertion procedure" and "None of the salt solution was administered before eye movements began."

Natural/Normal/Physiologic/Regular/Standard *adjective*

'Natural' means 'as occurs in nature without human intervention' as in: "Shampooing your dog washes out natural oils."

Use 'normal' with caution around numbers. Technically, 'normal' refers to a very specific grouping of numbers. 'Normal' is the name of the smooth bell-shaped curve fitting the equation $y = K \exp(-x^2/2)$, as used in: "The diameters of 344 hybrid peas have a normal distribution, with a mean of 0.83 mm and a standard deviation of 0.041 mm." To avoid confusion, don't describe numbers as normal in any other context.

'Physiologic' (or 'physiological') means 'within the range of a healthy functioning organism' as in: "To approximate the natural condition, we use physiologic doses of insulin to treat diabetes."

'Regular' means 'at even intervals (of either time or space)' as in: "The tones were regular, occurring once every 2.5 sec" or "The spots formed a regular line along its back—there was one spot every 4.2 mm."

'Standard' means 'matching a particular pre-defined description' as in: "We used the standard solution specified by Holtfreter (1953)" or "The apparatus was bathed in oxygen at body temperature (37°C) and standard pressure (760 mm Hg)."

Open/Opened *adjective or verb*

In the phrase 'to be open,' 'open' is an adjective. "The jar was open," reports simply that we found no cover on the jar.

In the similar-sounding phrase 'to be opened,' 'opened' is a verb. "The jar was opened," tells us:

(a) at one time, the jar had been covered,

(b) afterward, someone took off the cover.

In other words, "We discovered that the window was open," does not tell us whether the window had ever been closed. While "We discovered that the window was opened" tells us that at an earlier time, the window had been closed and that somebody then opened it.

Paradigm *noun*

'Paradigm' means 'model' or 'example' as in: "We used the Billington paradigm when designing our questionnaire." 'Paradigm' is not the same as 'archetype,' 'paragon,' or 'prototype.' These words mean 'original or particularly noteworthy models or examples' as in: "Recently, Billington's paradigm has replaced the previous prototype and is now considered to be the standard for designing questionnaires."

Pertinent/Relevant/Salient *adjective*

'Pertinent' means 'has a logical and precise bearing on the matter at hand' as in: "These are the two formulas that are pertinent to our geometric calculation."

'Relevant' is more general. It means only 'has a bearing on the subject at hand' as in: "A discussion of acute angles is relevant to our geometric calculation."

'Salient' is different. It means 'striking' and 'conspicuous' as in: "The reviewer's religious terminology was the most salient part of her commentary on our geometric calculation." Salient things need not be pertinent or relevant.

Random *adjective*

Technically, 'random' has two meanings.

(1) A machine or a process is random if its output is as unpredictable as possible. The classic example is coin flipping, a process that produces an unpredictable result each time it is tried.

(2) A sequence of numbers is random if its order is as heterogeneous as possible. This type of randomness is sometimes called "iid random," which stands for identically and independently distributed random.

When you are randomizing items in an experiment, you want to use iid random sequences of numbers. These sequences contain the maximum number of different subsequences or subpatterns. In this way, you make it unlikely that any patterns inherent in your experimental items will match the sequence that you assign them. When randomizing items in an experiment, you do not want to use numbers produced by a random process. If you flip a coin, you are not guaranteed to produce an iid random number. Instead, you may get a very orderly, homogeneous sequence, such as HHHTTT.

In your writings, use the word 'random' carefully. In most cases, it is better to use 'heterogeneous,' 'irregular,' 'disorganized,' or 'with no discernible pattern.' In cases where you have assigned items to an order without using a pre-chosen iid random sequence, you should say that the order was assigned 'arbitrarily' not 'randomly.'

Significant *adjective*

In a scientific paper, reserve this adjective for statistical descriptions, and then use it only when the appropriate statistical tests have been applied. Scientifically, 'significant' should always be followed by its mathematical limits, which are written as, "($p < 0.01$, t-test)." When using 'significant,' the sentence you write is actually shorthand. You write, "The difference between these two data sets was significant ($p < 0.01$, t-test)." But you mean, "Using the t-test, we have calculated that the chances are less than 1 out of 100 that these two data sets are just chance variants drawn from the same large set of possible data points."

That/Which *pronoun*

'That' separates out items from a group. "The drugs that caused an adverse reaction had bacterial contaminants." Here, of all the drugs under consideration, only those causing an adverse reaction are being discussed.

'Which' gives additional, parenthetical information about all the elements. "The drugs, which caused an adverse reaction, had bacterial contaminants." In the latter case, all the drugs under consideration caused an adverse reaction.

Via/Using *preposition*

'Via' means 'taking the route of.' For example, "We traveled from Maine to Florida via Route 1" or "We slipped the catheter in via the femoral vein."

'Via' does not mean 'by means of'—when you aim for this meaning, write, "We traveled from Maine to Florida using [not 'via'] a restored Model-T Ford," or "We injected glucose through the catheter using [not 'via'] a peristaltic pump."

Appendix B

SIMPLIFYING WORDY, REDUNDANT, AND AWKWARD PHRASES

Replace	With
a considerable amount of	many, much
absolutely essential	essential
almost unique	rare, uncommon
an order of magnitude more than	ten times
as to whether	whether
completely full	full
considered as	considered
considering the fact that	although, because
decline	decrease
different than	different from, unlike
due to the fact that	because
each and every	each
end result	result
equally as	equally
exact same, exactly the same	identical
exhibit a tendency	tend
final outcome	outcome
firstly, first of all	first
foregone conclusion	expected
foreseeable future	future
have a tendency	tend
having gotten	having got
help and	help to
higher in comparison to	higher than
if and when	if, when
in close proximity to	near
in spite of the fact that	although
in the final analysis	finally
in the realm of possibility	possible
including but not limited to	including

189

inside of	inside
intimate	suggest, indicate
irregardless	regardless
last but not least	finally
methodology	method, methods
multiple	many
nearly unique	rare, uncommon
obviate	prevent
orientate	orient
preventative	preventive
prove conclusively	prove
referred to as	called
regarded as being	regarded as
seeing that	because
the question of whether	whether
transpire	happen
try and	try to
up in the air	undecided
very unique	unique
whether or not	whether

Appendix C

STANDARD SCIENTIFIC ABBREVIATIONS

Basic Rules

- Use metric units, such as meters, kilograms, and liters
- Use Celsius temperatures, not Fahrenheit
- Arterial blood pressures are reported in millimeters of mercury (i.e., "120 mmHg")
- Unless you are using one of the standard abbreviations listed below, define the abbreviation when it is first used in the paper (e.g., "nerve growth factor (NGF)")
- Species names should be italicized (e.g., *Homo sapiens*)—after its first usage in a manuscript, the genus can be replaced by its first letter (e.g., *H. sapiens*).
- Genes, mutations, genotypes, and alleles should be indicated in italics. Use the recommended name by consulting the appropriate genetic nomenclature database—for human genes the names can be found at HUGO (http://www.genenames.org/index.html).
- Gene prefixes, such as those used for oncogenes or cellular localization, should be printed in regular type, such as, v-fes and c-MYC.

Standard Abbreviations
- Standard abbreviations are used only after a number, for example, "4 atm" but "many atmospheres"—unless specifically indicated below.
- Standard abbreviations are used without a period, for example, "4 atm" not "4 atm."—unless specifically indicated below.
- Standard abbreviations are the same in the singular and the plural, for example, "4 atm" not "4 atms"—unless specifically indicated below.

ampere   A
atmosphere   atm

byte   B

calorie   cal
candela   cd
Celsius   C
centigrade   C
centimeter   cm
counts per minute   cpm
Curie   Ci

cycles per second   cps

day   d
decibel   db *or* dB

gigabyte   GB
gram   g

hertz   Hz
hour   h

intravenous   i.v. [*does not need to follow a number*]

Kelvin   K
kilobase   kb
kilobyte   kB
kilocalorie   kcal
kilogram   kg
kilohertz   kHz
kilometer   km

liter   l
logarithm (base 10)   log
logarithm (base *e*)   ln

megabyte   MB
megahertz   MHz
meter   m
microgram   μg
microliter   μl
micrometer   μm
micron   μm
milliampere   mA
millicurie   mCi
milligram   mg
milliliter   ml
millimeter   mm
millimole   mmol
millirad   mR
millisecond   ms *or* msec
minute   min
molar, moles/liter   M
mole   mol
molecular weight   mol wt *or* MW
month   mo

nanogram   ng
nanometer   nm
number   no.

parts per million   ppm
picogram   pg
pounds per square inch   psi

rad   R
radian   rad
revolutions per minute   rpm

second   s
species (*singular*)   sp. [*does not need to follow a number*]
species (*plural*)   spp. [*does not need to follow a number*]
specific gravity   sp. gr.
standard deviation   SD

unit   U

volts   V
volume   vol

watt   W
week   wk
weight   wt

year   y *or* yr

Details on the International System of Units (SI units) can be found at:
http://www.bipm.org/en/si/

Appendix D

TYPICAL BIBLIOGRAPHIC FORMATS

When you choose a journal, you will probably have to do some fiddling to adjust your *References* and your text citations to match the appropriate format. If you have been using bibliographic software, these adjustments should be fairly painless.

1. FORMATS FOR TEXT CITATIONS

All items in your *References* section must be cited in your paper. The form of the text citations varies from journal to journal; for example:

* Some journals ask you to number your *References* in the order in which they are first mentioned in the text of your paper. Other journals ask you to alphabetize your *References* and to refer to them by authors' last names in the text (e.g., "(Katz and Lasek, 1985b)").
* Some journals ask you to cite full online references in parenthesis in the text. Other journals ask you to number online references and put the full citation in the *Reference* section.

2. BIBLIOGRAPHIC FORMATS FOR THE *REFERENCES* SECTION

The exact format of the full bibliographic information about your reference materials also varies from journal to journal. The *International Committee of Medical Journal Editors* has been working to standardize bibliographic formats in the biomedical sciences. Their recommendations are published on the National Library of Medicine website: *http://www.nlm.nih.gov/bsd/uniform_requirements.html*

Here are some of the basic formats from that website:

I. Articles

A. Print
   1. Standard article
Halpern SD, Ubel PA, Caplan AL. Solid-organ transplantation in HIV-infected patients. N Engl J Med. 2002; 347(4): 284–7.

2. Standard article with >6 authors

Rose ME, Huerbin MB, Melick J, Marion DW, Palmer AM, Schiding JK, et al. Regulation of interstitial excitatory amino acid concentrations after cortical contusion injury. Brain Res. 2002; 935(1–2): 40–6.

3. Nonstandard article (e.g., letter or commentary)

Tor M, Turker H. International approaches to the prescription of long-term oxygen therapy [letter]. Eur Respir J. 2002; 20(1): 242.

4. Article published online before it is in print

Yu WM, Hawley TS, Hawley RG, Qu CK. Immortalization of yolk sac-derived precursor cells. Blood. 2002; 100(10): 3828–31. Epub 2002 Jul 5.

5. Article accepted and pending publication

Tian D, Araki H, Stahl E, Bergelson J, Kreitman M. Signature of balancing selection in Arabidopsis. Proc Natl Acad Sci U S A. In press 2002.

B. Online article

Abood S. Quality improvement initiative in nursing homes: the ANA acts in an advisory role. Am J Nurs [serial on the Internet] 2002 Jun [cited 2002 Aug 12]; 102(6): [about 3 p.]. Available from: *http://www.nursingworld.org/AJN/2002/june/Wawatch.htm*

II. Books

A. Print

1. Authored book

Murray PR, Rosenthal KS, Kobayashi GS, Pfaller MA. Medical microbiology. 4th ed. St. Louis: Mosby; 2002.

2. Edited book

Gilstrap LC 3rd, Cunningham FG, VanDorsten JP, editors. Operative obstetrics. 2nd ed. New York: McGraw-Hill; 2002.

3. Chapter in a book

Meltzer PS, Kallioniemi A, Trent JM. Chromosome alterations in human solid tumors. In: Vogelstein B, Kinzler KW, editors. The genetic basis of human cancer. New York: McGraw-Hill; 2002. p. 93–113.

B. Online Book

Foley KM, Gelband H, editors. Improving palliative care for cancer [monograph on the Internet]. Washington: National Academy Press; 2001 [cited 2002 Jul 9]. Available from: *http://www.nap.edu/books/0309074029/html/*

III. Other

A. Print

1. Conference paper

Christensen S, Oppacher F. An analysis of Koza's computational effort statistic for genetic programming. In: Foster JA, Lutton E, Miller J, Ryan C, Tettamanzi AG, editors. Genetic programming. EuroGP 2002: Proceedings of the 5th European Conference on Genetic Programming; 2002 Apr 3–5; Kinsdale, Ireland. Berlin: Springer; 2002. p. 182–91.

2. Dissertation

Borkowski MM. Infant sleep and feeding: a telephone survey of Hispanic Americans [dissertation]. Mount Pleasant (MI): Central Michigan University; 2002.

3. Dictionary or similar reference book

Dorland's Illustrated Medical Dictionary. 29th ed. Philadelphia: W.B. Saunders; 2000. Filamin; p. 675.

B. Online webpage

Cancer-Pain.org [homepage on the Internet]. New York: Association of Cancer Online Resources, Inc.; c2000-01 [updated 2002 May 16; cited 2002 Jul 9]. Available from: *http://www.cancer-pain.org/*

Appendix E

ADDITIONAL READING

There is much good advice on scientific writing available in libraries and on the Internet.

Books about technical writing
- Beall H, Trimbur J. 2001. A Short Guide to Writing About Chemistry, 2nd ed. Longman, New York.
- Day RA, Gastel B. 2006. How to Write and Publish a Scientific Paper, 6th ed. Greenwood, Westport, CT.
- Huck SW. 2007. Reading Statistics and Research, 5th ed. Allyn & Bacon, Boston.
- Zobel J. 2004. Writing for Computer Science, 2nd ed. Springer, New York.

Standard style manuals
- The Chicago Manual of Style, 15th ed. 2003. Univ Chicago Press, Chicago.
- Siegal AM, Connolly WG. 2002. The New York Times Manual of Style and Usage. Three Rivers Press, New York.

Style manuals specifically for scientific writing
- Coghill AM, Garson LR. 2006. The ACS Style Guide: Effective Communication of Scientific Information. Am Chem Soc, Washington, DC.
- The American Medical Association Manual of Style: A Guide for Authors and Editors, 10th ed. 2007. Oxford Univ Press, New York.
- Scientific Style and Format: The CSE Manual for Authors, Editors, and Publishers, 7th ed. 2006. Council of Science Editors, Reston, VA.

Guide to statistics
- Freedman D, Pisani R, Purves R. 2007. Statistics, 4th ed. W.W. Norton, New York.

Guide to scientific figures
- Tufte ER. 2001. The Visual Display of Quantitative Information, 2nd ed. Graphics Press, Cheshire, CT.

Appendix F

SOFTWARE SUGGESTIONS

Bibliographic & Reference Records
- *Zotero* (Center for History and New Media, George Mason University) [can be downloaded from: *http://www.zotero.org/*]

Graphing
- *Origin 7.5* or later (OriginLab)

Lab Notebook/Diary
- *EverNote* (EverNote Corp) [can be downloaded from: *http://www.evernote.com/*]

Photo Editing
- *Photoshop Elements* (Adobe)

***Statistical* (Exploratory)**
- *JMP* (SAS)

Wordprocessing
- *Word* (Microsoft)

Index

LaVergne, TN USA
16 December 2010
209003LV00002B/4/P

9 781402 094668

[7